QC791.775.C64 T38 1993
Taubes, Gary
Bad science

3 4569 00077288 7

Gary W9-BWM-895

Bad science

DATE DUE		
MAY 1 2 1995 S		
JUN 2 6 1995 S		
DEC 1 3 1996 S		

WITHDRAWN

ALSO BY GARY TAUBES

Nobel Dreams

BAD SCIENCE

BAD SCIENCE

The Short Life and Weird Times of Cold Fusion

GARY TAUBES

RANDOM HOUSE

NEW YORK

Copyright © 1993 by Gary Taubes

All rights reserved under International and Pan-American Copyright Convention.
Published in the United States by Random House, Inc., New York.

Library of Congress Cataloging-in-Publication Data

Taubes, Gary
Bad science: the short life and weird times of cold fusion / Gary Taubes.
p.   cm.
ISBN 0-394-58456-2
1. Cold fusion. I. Title.
QC791.775.C64T38 1993
539.7′64—dc20   91-52693

Manufactured in the United States of America
24689753
First Edition

For my parents,
Ernest Paul Taubes
and
Zelda Taubes (April 20, 1920, to May 30, 1991)

299906

306.08

The rare genius with a flair for research will not benefit from instruction in the methods of research, but most would-be research workers are not geniuses, and some guidance as to how to go about research should help them to become productive earlier than they would if left to find these things out for themselves by the wasteful method of personal experience.

WILLIAM BEVERIDGE,
The Art of Scientific Investigation

The first principle is that you must not fool yourself and you're the easiest person to fool.

RICHARD FEYNMAN,
"Surely You're Joking, Mr. Feynman!"

AUTHOR'S NOTE

The cold-fusion episode teaches two lessons that can be applied as meaningfully to journalism as to science:

1. Do your research, because nothing is as simple as it seems.
2. Make sure you've got the story right before you publish.

With these points in mind, over 260 persons were interviewed for this book between March 27, 1989, and November 1992. Some of the interviews were as brief as a five-minute phone conversation; most took hours or even days. Quite a few sources kindly consented to repeated telephone calls and time-consuming interviews for the duration of the three-year period. A list of those interviewed can be found at the back of the book. Conspicuously absent from this list are Stan Pons and Martin Fleischmann, who refused numerous requests for interviews. Martin Fleischmann, however, was kind enough on several occasions to answer a few brief questions before having second thoughts.

The book was read in draft form and corrections suggested by Allen Bard, Tim Fitzpatrick, Richard Garwin, William Happer, John Huizenga, Steven Koonin, Del Lawson, Nathan Lewis, Charles Martin, Hugo Rossi, and Michael Salamon. Any errors remaining in either fact or form, however, are mine alone.

A regrettable side effect of interviewing so many individuals on such a small, however perverse, subject is that one invariably ends up with considerably more information than can be accommodated in a book of reasonable size. My editors, justifiably, refused to publish 700-plus pages on cold fusion. As a compromise, they allowed me to transfer much of the less pertinent information to the end notes. This section of the book provides a second or even third level of information, speculation, perspective, and humor into which readers can delve if they choose.

CONTENTS

*O*n March 16, 1989, University of Utah president Chase Peterson agreed with his scientists and lawyers that it was no longer possible to keep secret the discovery of cold nuclear fusion. As he later explained, the determining factor in his decision was not that his scientists had unambiguous evidence of this remarkable breakthrough, only that such a nuclear reaction might be possible and that they might have created one in their basement laboratory.

The research had been the brainchild of B. Stanley Pons, chairman of the Utah chemistry department, and Martin Fleischmann, Pons's British collaborator and mentor. They had told Peterson that they had been working together on cold nuclear fusion for five years and that they needed eighteen more months to complete their experimental protocol. But by March 1989, events beyond their lab were dictating the schedule of their research. Time had become a luxury they could not afford. At nearby Brigham Young University, a physicist named Steven Jones had come upon the same remarkable phenomenon. Pons and Fleischmann believed that Jones had not discovered cold fusion independently but had "pirated" the key ideas from them.

Six months earlier, the two chemists had submitted a proposal to the Department of Energy requesting funding for their cold fusion experiments. The DOE had sent the proposal on to Jones for review. Pons and Fleischmann believed that Jones had read their proposal and then run a quick and dirty version of *their* experiment. Now Jones not only claimed that he had equal right to pursue the research but within a month would disclose that *he* had discovered cold fusion.

On March 16 Pons and Fleischmann had virtually no data to support their experimental conclusion—which is to say, *their* discovery of cold nuclear fusion—but because of this untimely challenge they had no choice but to make a public announcement and to do it quickly. If not, they would lose credit for what appeared to be one of the great scientific discoveries of all time. The chemists and the university administration agreed to schedule a press conference for March 23, 1989.

Until then, secrecy would be paramount. Stan Pons informed his own researchers of the press conference only two days in advance and swore them to silence on the few details he was willing to divulge. By then they knew that something big was in the works because Pons had seemed even more frenetic than usual. His working hours, which had always been long, had lately verged on twenty-four hours a day. And what the graduate students called "administration types" had begun traipsing mysteriously into the back room where the fusion cells were hidden.

Pam Fogle, news director at the University of Utah, only began alerting her contacts in the media on March 22. Fogle had been warned by her boss, a fellow with twenty-five years' experience in public relations, that the discovery was so astonishing that reporters were bound to break any embargo if given half a chance. Following this strategy, Fogle did not send out press releases in advance, fearing they might "fall into the wrong hands." To some reporters—Philip Hilts of *The Washington Post,* for instance—Fogle refused even to divulge what the announcement would be, only that it was tremendously important and that it would be made at 1:00 P.M. the next day. This sort of secretive argument didn't sway Hilts, so he didn't make the trip to Salt Lake City.

Another of Fogle's first calls went to Jerry Bishop, the veteran science reporter of *The Wall Street Journal.* Fogle needed the *Journal*'s coverage, so with Bishop she risked revealing her news—that the University of Utah had achieved a sustained nuclear fusion reaction at room temperature for one hundred days.

"That's ridiculous," he said, "but thanks for letting me know." Bishop then went off to lunch with a science writer, who suggested that maybe it wasn't so ridiculous, and he set out to track down the details.

Despite Fogle's precautions, the CBS News affiliate in Salt Lake City broke the news of cold nuclear fusion that night, and *The Wall Street Journal* and the *Financial Times* of London weighed in the next morning. Rather than diminish interest in the press conference, however, these leaks further legitimized the university's announcement.

With the news spreading fast, campus police appeared in force at Pons's basement laboratory on the morning of the twenty-third to guard the discovery of the century. A half-dozen uniformed officers and at least as many plainclothes detectives stationed themselves at strategic points in and around the lab. Security was so tight that when the detectives realized one of the laboratory doors had a window, they ordered the students to cover it with black paper. The graduate students found these actions ludicrous.

Inevitably, rumors began spreading throughout the lab and the university at large. One had it that President George Bush would attend the announcement; then it was Margaret Thatcher who was coming. Finally, it was whispered that Vice President Dan Quayle would be there, although he didn't show either. Considering the magnitude of the discovery, no rumor seemed too outrageous to be rejected out of hand.

The second day of spring in Salt Lake City was a balmy one, with a touch of winter lingering in the air. The mountains behind the city and the campus were capped with snow. Many of the faculty and students of the University of Utah would spend the weekend skiing.

By 1:00 P.M., the lobby of the Henry B. Eyring Chemistry Building was overflowing. Visitors were packed half a dozen deep by the rear doors. Seven television cameras stood on tripods along the back wall; klieg lights offered a hint of the surreal.

The cherubic Jim Brophy, with his white hair and rosy cheeks, vice president for research at the university, began the press conference and made the appropriate introductions. He started with Robert Nesbitt, a dean in the Faculty of Science at the University of Southampton, England. Nesbitt had flown over the previous night to represent, in Peterson's words, "the international cooperation in scientific exploration" that had made the discovery possible.* Sitting next to him was Chase

---

*This was the only role Nesbitt would play, and his brief appearance would be the last anyone would hear of the spirit of international cooperation throughout the cold-fusion affair.

Peterson. And to Brophy's right were Stan Pons and Martin Fleischmann.

"We are here today," Peterson intoned, by way of introduction, "to consider the implications of a scientific experiment." A nonpracticing physician trained at Harvard and Yale, Peterson projected the avuncular bedside manner of a family doctor. He was already well known to science reporters. Six years earlier, when University of Utah surgeons had implanted the first artificial heart in Barney B. Clark, Peterson had run the show. Reporters had found Peterson dignified, charming, trustworthy, and very much in control. *The New York Times* had called him "a voice of calm in periods of high excitement and repeated crisis."*

As he introduced the discovery of cold fusion, Peterson's manner was so understated that he might have been announcing the hiring of a new professor or unveiling plans for a new medical center. But his even tone was overwhelmed by the flagrant optimism of his message. Over the past several decades, Peterson explained, the U.S. government had spent some $8 billion pursuing research in nuclear fusion. It was considered the energy source that would save humankind: the mechanism that powers the sun and stars, harnessed to provide limitless amounts of electricity. Since shortly after the Second World War, physicists had worked to induce, tame, and sustain fusion reactions by re-creating the hellish heat and pressure at the center of the sun in a controlled setting. The conventional wisdom was that sustained nuclear fusion could only be achieved in the laboratory with enough heat—tens of millions of degrees—and extraordinary technological wizardry. Every industrialized nation in the world, at one time or another, had attempted to create and sustain nuclear fusion, but progress was slow. Physicists were still decades away from creating a viable fusion reactor.

Now, Peterson was saying, these physicists might be out of business. The accompanying press release would put it in even more confident terms: "Two scientists have successfully created a sustained nuclear fusion reaction at room temperature in a chemistry laboratory at the University of Utah. The breakthrough means the world may someday rely on fusion for a clean, virtually inexhaustible source of energy." What more was there to say?

When Peterson finished his remarks, Stan Pons rose to speak. The forty-six-year-old chemist was wearing a conservative dark blue suit and a polka-dot tie and appeared pale. His hair was cut in a peculiar style that

---

*Coincidentally, the announcement of cold fusion fell on the sixth anniversary of Barney Clark's death.

The Boston Globe would later liken to a Buster Brown and the *Los Angeles Times* to a Julius Caesar. There was something about him—perhaps the hair or his nervous manner or his thick eyeglasses—that made him seem like the archetypal scientific egghead. In his hands, Pons cradled a model of his cold fusion reactor, which looked like—and was—nothing more than a sophisticated test tube about the size of a highball glass.

Pons cleared his throat and looked down at his notes; he spoke quickly and so quietly that his soft southern drawl could barely be heard. He thanked the university for its support and Martin Fleischmann, his codiscoverer, of whom he said, "a finer scientist and person would indeed be hard to find." Then he said, "We've established a sustained nuclear reaction by means which are considerably simpler than conventional techniques." In fact, as Pons explained, their laboratory setup was similar to what might be found in a freshman-level college chemistry course.

Then Martin Fleischmann stood, took the reactor from Pons, and explained that cold nuclear fusion could be sustained "indefinitely." Fleischmann, age sixty-one, was visiting from the University of Southampton, where he had built a name for himself as one of the most distinguished electrochemists in the world. His thinning brown hair was combed over the crown of his head, and he wore dark-rimmed eyeglasses. He stooped slightly and spoke with a clipped and very distinctive Slavic brogue that betrayed his Czechoslovakian birth and English upbringing. Fleischmann seemed tired and ill at ease that afternoon. Of the five men at the table, he may have been the only one who fully grasped the magnitude of their actions. Just that morning he had warned one of the graduate students that their lives would never be the same after they went public; the response, he had said, would be astounding.

To the assembled reporters, Fleischmann described the nature of the experiment that had achieved such remarkable results. In nuclear fusion, he explained, lighter atoms join together—or fuse—to form heavier atoms. In particular, he and Pons had managed to generate the fusion of deuterium atoms, which are a heavy form of hydrogen, by compressing them inside their cold fusion cells. These cells consisted simply of two metal electrodes—one palladium and one platinum—connected to a moderate electric current and submerged in a bath of heavy water (in which hydrogen atoms have been replaced by deuterium atoms). To this was added a dash of lithium. "That," he said, "is really the guts of the experiment."

They had achieved this stupendous breakthrough, Fleischmann said, on a shoestring budget: "We thought this experiment was so stupid that we financed it ourselves." He even showed a slide of a cold fusion reactor mounted in a dishpan. This was for lecture purposes only, Fleischmann explained, to demonstrate that they "couldn't actually pay for very much so it had to be done in a Rubbermaid basin." Even so, they "had burned up about a hundred thousand dollars" of their own money, which might sound like a lot, but not compared with the tens of billions spent by physicists trying to achieve the same result.

For the next thirty minutes, Pons, Fleischmann, Peterson, and Brophy answered questions from the reporters and scientists. The proceedings were subdued by the standards of a political press conference, orderly and quiet. There was no shouting or interruptions. None of the reporters knew quite what to ask. Nuclear fusion, after all, is a complicated process. What evidence did they have for this sustained fusion reaction? A miniature sun, perhaps, in their basement?

In a few short years, Pons and Fleischmann said, they should be able to build a fully operational nuclear fusion reactor that could produce electric power. They said that the oceans of the world would supply fuel for these reactors for millions of years. Although no one explicitly said that science and technology, once again, had saved the world, that belief was implicit in their message. In the weeks before the press conference, Peterson, at least, had envisioned what limitless, clean energy would mean to a world relying on fossil fuels, a world with serious energy and environmental problems. Not only would it eliminate acid rain and the dangers to the ozone layer but it would negate the United States' dependency on Middle Eastern oil. There would be no more Three Mile Islands or Chernobyls, no more endless accumulation of hazardous nuclear waste. And, of course, cold fusion represented the potential for making huge sums of money—equal to the wealth of OPEC at least. There were billions and billions of dollars to be made and Nobel Prizes to be won. Indeed, cold fusion seemed like salvation in more ways than one.

Peterson read a congratulatory letter from Norman Bangerter, the governor of Utah, in which he said that cold fusion proves, once again, that "this is the place"—echoing the words of Brigham Young, who with his Mormon followers had arrived in this promised land in 1847.

As the press conference concluded, however, Peterson may have had a momentary lapse in confidence. He leaned over to Brophy, who had been trained as a physicist, and asked whether he truly believed that Pons and Fleischmann had discovered cold nuclear fusion. Brophy replied that he had no doubt at all. It was all Peterson needed to reinforce his faith. He told reporters that their discovery "ranks right up there with fire, with cultivation of plants, and with electricity." Brophy added that Pons and Fleischmann had reproduced their experiment and confirmed their conclusion dozens of times. The data, he said, were "overwhelming."

By the following morning, cold fusion had made front-page news throughout the world, and it would continue to do so for months. Over the next year the scientific community would spend nearly $100 million trying to confirm the Utah discovery, and cold fusion would become the most controversial, if not the most bizarre, scientific episode in decades. If after the press conference Pons, Fleischmann, Peterson, or Brophy considered the possibility that they might be dead wrong about the existence of cold fusion, they rarely, if ever, let it show.

Book I

DELUSION

IS THE BETTER

PART OF

GRANDEUR

Anyone with an alertness of mind will encounter during the course of an investigation numerous interesting side issues that might be pursued. It is a physical impossibility to follow up all of these. The majority are not worth following, a few will reward investigation and the occasional one provides the opportunity of a lifetime. How to distinguish the promising clues is the very essence of the art of research.

WILLIAM BEVERIDGE,
The Art of Scientific Investigation

The only way of making clear pea soup is by omitting the peas.

A. J. LIEBLING,
The Press

*I*n the March 23, 1989, press release announcing the discovery of cold fusion, Stan Pons noted that the odds of successfully generating a nuclear fusion reaction in a test tube were a billion to one against. He also said that the experiment, in theory, "made perfectly good scientific sense." Thus he and Martin Fleischmann had initiated their paradoxical exercise in nuclear research "for the fun of it" and to satisfy scientific curiosity. Once they performed the experiment, however, they had "immediate indication that it worked."

When asked at the press conference what this evidence was—"when you discovered that this one chance in a billion comes through," as a reporter phrased the question—Fleischmann told how they had induced what appeared to be a nuclear meltdown in a cube of palladium. "The thing which really triggered the whole thing off fairly early on," Fleischmann said, "was that we realized that you could generate a lot of heat, a lot."

For Pons and Fleischmann, "the meltdown" or "the explosion," as it

3

came to be called, was the pivotal event in their research. It most likely occurred in the autumn of 1984 or the winter of 1985 and represented the dramatic culmination of Pons and Fleischmann's very first, primitive experiment in cold fusion. They had suspended a solid, one-cubic-centimeter palladium electrode from a palladium wire in a large beaker. The beaker was filled with a cocktail of heavy water and lithium. They charged the contraption by passing a current between the palladium electrode and a second electrode (if their future experiments were any indication) made of platinum.

They charged the cell for seven months until one fateful evening when young Joey Pons, who had been running the experiments for his father, lowered the current and left for the night. Hugo Rossi, the dean of Utah's College of Science and the first director of the National Cold Fusion Institute, later described Joey Pons as having the appearance of a simple farm boy and said it was hard to believe that his father would trust him in the laboratory with nuclear research. But Joey was family, which counted for quite a bit with Stan Pons. The meltdown occurred during the night. There were no witnesses.

Joey must have discovered the evidence the next morning, because he went to his father and asked him to take a look. Pons found that the palladium cube had melted and partially vaporized. It may also have evaporated the cell contents and melted through the beaker, destroying the experiment and a protective enclosure. It may have blown the apparatus apart. The accounts differ.[1]

Fleischmann, who rarely missed the opportunity to couch his research in the technical jargon of nuclear physics, explained that their primitive cell had reached "ignition," the point at which enough heat is generated by the fusion reactions to make the reactor self-sustaining. As Fleischmann told it, they had a "runaway," but they managed to "contain" it. The cell, he said, had no "failsafe."

When Pons described the incident to *The Wall Street Journal,* he added his own dramatic flourishes. In this account, they increased, not decreased, the current on the apparatus. The exploding, or melting, or vaporizing cube of palladium heated up to thousands of degrees and burned a four-inch-deep hole in the lab's concrete floor. (Kevin Ashley, a graduate student working for Pons at the time, later described the hole as truly impressive. "I was in the lab the next morning. There was this huge goddamn hole in the floor and through the counter." Ashley, however, did not realize that the damage had been wrought by an attempt to generate nuclear fusion in a test tube. He had assumed only that "it was some weird stuff Stan was doing on the side [for the Office of Naval Research].")

The additional details provided by Pons prompted quite a few scientists to attempt to calculate the thermodynamics of the system: how much energy is required to melt a cube of palladium and then burn through four inches of concrete? Could a chemical reaction do this, or only a nuclear reaction? They were hampered, however, by lack of information. Was it a crater, or just a hole? And if the hole was four inches deep, how wide was it?

The hole's appearance in press accounts also raised some perplexing questions about how laboratory research was conducted at Utah. To Marcel Gaudreau, for instance, a nuclear engineer at the Massachusetts Institute of Technology, the hole brought up an obvious problem. "If you had a hole in the floor in the concrete at MIT," he observed, "there'd be a safety investigation; you'd have the campus fire department there. If we had a hole in the floor, no matter what it was from, it would be investigated." And Stephen Feldberg, an electrochemist at Brookhaven National Laboratory, later made the same point. If the safety engineers at Brookhaven had been told of such an explosion after the fact, he said, "they would have wet their pants."

All of this, regrettably, begged the question of whether the meltdown had released radiation. It is the nature of nuclear reactions that they blow off energetic bits and pieces—neutrons, protons, electrons, alpha particles—of the nuclei involved or radiate away the energy released in the guise of gamma rays or esoteric particles known as neutrinos. Pons and Fleischmann believed they had induced the fusion of deuterium nuclei, which invariably results in the emission of neutrons, among other things.[2] If deuterium fusion were to account for the meltdown of a cube of palladium, then the fusion cell itself should have emitted trillions of neutrons per second. This would have constituted a serious, if not lethal, dose of radiation.

The Wall Street Journal of April 3, 1989, quoted Pons saying only that the explosion had resulted in "a nice mess." To this the report added, "A check of the laboratory the next day with a radiation counter indicated radioactivity levels three times higher than the normal background levels, apparently the result of a sudden spray of neutrons. The radioactivity, [Pons] said, was far below any dangerous level." This account had been contradicted by Pons three days earlier, when he told several hundred of his colleagues at the University of Utah that they had detected no radiation. Three days before that he had told Allen Bard, a distinguished chemist at the University of Texas, that the meltdown had left the laboratory "contaminated." It was anybody's guess.

Fleischmann reported on March 31 at a lecture in Geneva, Switzerland, that the meltdown had resulted in "no untoward radiation about" and that after the accident they had redesigned the experiments "to be absolutely safe." He explained that they had switched from cubes to thin sheets and later rods of palladium and thus had been slowly working back toward conditions that would be ripe for "ignition." That was one reason, he said, why the research had taken so long to come to fruition.

Although Pons and Fleischmann had written over thirty scientific papers together, there were those in Utah who would say that their most celebrated collaborative effort before March 23, 1989, was their annual Christmas party. Fleischmann would work with Pons in Utah for several months, then return in December for their annual soiree. The two would spend weeks buying and preparing the food, which they would lay out in dishes with elaborately scripted labels. Pons would put up an eighteen-foot Christmas tree in the vaulted living room of his Salt Lake City home. A string quartet would play. It was quite a bash, and even Pons's funding agents from Washington would appear to join the celebration.

It may have been this kind of continual, close, friendly teamwork that brought these scientists together for their remarkable collaboration. As the two would later tell reporters, they both liked to ski, hike, cook, and drink. Fleischmann would say that "Stan and I do a lot of cooking together" and that cold fusion had been cooked up in the course of their culinary sessions.

The two chemists had met in 1975, when Pons arrived to finish his graduate studies at the University of Southampton in England, where Fleischmann was a senior professor. Pons liked to tell how he first saw Fleischmann, the distinguished chemist, in the midst of an experiment demonstrating "the high level of sophistication that electrochemical experimentation had attained." At the time, Pons was a North Carolina boy who had studied at Wake Forest University and had dropped out of the University of Michigan two years into his graduate work. After working in his family business for ten years, he returned to his studies at the University of Southampton, where he found one of the most prestigious electrochemists of his day "clearing the corridors outside the laboratories in preparation for a high velocity extrusion experiment. One graduate student was about to launch a molten glass arrow from a cross bow. The molten glass contained a lead wire that would soon have its diameter rapidly decreased while having its length correspondingly increased." Pons "was now absolutely sure that he had made the proper choice of graduate schools (well, almost)."[3]

From then on Pons saw the elder British chemist as a friend, a father figure, and a mentor. And Fleischmann, with three children of his own, seemed to treat Pons as a son. They made quite a pair, the younger, aggressive American and the older, worldly Englishman. Pons was just beginning his career in chemistry. Fleischmann was on the verge of retirement after a long and distinguished career. Pons was the experimentalist. Fleischmann was the theorist.

Pons had a small-town sensibility. He hailed from Valdese, a town of 3,000 or so in the foothills of the Blue Ridge Mountains. Follow Route 40 west from Hickory to Asheville and you'll pass Valdese, deep in the heart of hillbilly country. The town, as the name implies, had been settled at the turn of the century by Waldenses, descendants of Italian Protestants who had suffered through a particularly bloody 800 years of persecution at the hands of the Roman Catholic church. They had bequeathed to Pons, if nothing else, his Mediterranean features, the aquiline nose, prominent eyebrows, and thick, now graying hair.

Pons would describe his youth as having been as mundane and American as could be. He was the typical kid with a chemistry set, running track in high school, playing the saxophone, and chasing girls. By the time cold fusion came along, he was on his third marriage. Friends of Stan Pons would say that they had never met a finer, nicer man. He was not the least bit flamboyant, but thoughtful, with an infectious, almost childlike enthusiasm. He seemed to be an honest-to-God, polite good ol' boy. He was quiet, modest, and self-effacing: Pons would talk about his own work as though it were trivial, even though he had made, with Fleischmann, what his peers considered major contributions to micro-electrode research.

And Pons was memorably generous. No one could recall going to dinner with him and paying, even if there were twenty people at the table. Invariably Pons picked up the check in such a way that no one could even argue. There was the time a group of his chemist friends drove down to Caesar's in Tijuana after a meeting in San Diego, drank uncountable margaritas and ate everything on the menu. When they asked for the check, the waiter apologized and said it had already been paid. Some of the younger chemists said they would have liked to emulate Pons's generosity, but then he had money and they didn't. The story was that his family had made millions in the textile industry, but no one knew for sure.

He was known among his peers as a gadget man. The more technological wizardry it took to pull off a piece of research, the more Pons seemed to like it. He even operated an electronics equipment company—selling his own brand of "fine electrochemical and spectroelectrochemical products"—out of his home in the middle-class neighborhood of Sugarloaf in Salt Lake City.

Pons had also been remarkably productive as a chemist. From the moment he returned to his graduate studies in 1975, he'd seemed intent on making up for his lost decade. He had finished his doctorate at Southampton in only two years, then moved from one teaching post to the next. He spent two years at Oakland University in Michigan, then three at the University of Alberta. He requested several times that the Alberta administrators consider accelerating the tenure process on his account. He felt that a man in his mid-thirties should already have tenure. When the chemistry department refused to make haste with his promotion, Pons went hunting for new opportunities and found one at Utah. He spent three years at Utah before becoming a full professor in 1986. Pons's publications went from one per year at Southampton to ten at Alberta, twenty as an assistant professor at Utah, then thirty-six, in 1988. Thirty-six published papers in one year comes to one paper every ten days.

Fleischmann celebrated his sixty-second birthday six days after the announcement of cold fusion. An urbane European, Fleischmann had been raised in England after he and his family fled the Nazi occupation of Czechoslovakia in 1939. His career in chemistry had been long and distinguished. He studied physical chemistry at Imperial College and received his doctorate from the University of London. He taught at Newcastle University for sixteen years before accepting the chair of electrochemistry at Southampton in 1967. Southampton had the most powerful electrochemistry group in the world, and Fleischmann was considered its driving force. He had won virtually every prize an electro-chemist could win, and in 1985 he was elected a fellow of the Royal Society, the highest scientific accolade the United Kingdom has to offer; many consider it only one step down in the hierarchy of scientific recognition from the Nobel Prize.

Although Fleischmann had technically retired in 1983, he had been doing more research than ever, splitting his time among Southampton, Pons's laboratory in Utah, and the Harwell laboratory, west of London. Fleischmann observed that this arrangement allowed him to diffuse his interests to the point where he usually had a dozen projects going at once. He cheerfully described himself as a "person without responsibility."

Many of his peers considered Fleischmann a rare breed in chemistry, possessed of an original and highly imaginative intellect. He knew his field cold and also could think quantitatively; he could take a complex physical problem and make it easy to grasp by formulating it mathematically. Fleischmann liked to call himself an applied mathematician who knew a lot of chemistry, and he obviously knew a great deal of physics as well. When Chase Peterson and the Utah administration were deciding how to handle cold fusion, Fleischmann's apparent facility with physics helped convince them of the experiment's credibility.

Fleischmann had a prolific, almost reckless, imagination. He could walk into a lab, be informed of a problem, and in ten minutes suggest ten ways to solve it. Nine out of the ten ideas might be wrong, but the tenth would at least be worth trying. Pons's graduate student Marvin Hawkins said that this imagination made being around Fleischmann stimulating. "You meet [Fleischmann]," he said, "and you're going, 'Wow, if I can spend just a month with this guy, I'll be a god myself.' "

When Pons and Fleischmann discussed science, Hawkins added, there was no doubt that Fleischmann was the dominant figure. "It's like a family gathering," Hawkins said. "Everybody throws their six cents in, but Dad directs the discussion."

Fleischmann's intellect was so fertile that he had rarely stayed with one piece of work for long. His career had been characterized by original entrées into a field employing a potentially revolutionary idea; then, while his colleagues were hurrying to catch up, he would move on to something new. Other chemists spent careers developing ideas that Fleischmann had introduced. He had been one of the first electrochemists to realize that there were better ways to analyze electrochemical phenomena than simply measuring currents and voltages and thus had helped generate a renaissance in the field. In the 1960s Fleischmann had turned to various forms of spectroscopy, from X rays to infrared, to see what he could learn about the reactions occurring on the surface of electrodes.

It was this line of work that led to what chemists offered as the primary evidence of Fleischmann's genius. It involved a technique called Raman spectroscopy and was one of those ideas about which the prevailing wisdom was that it never should have worked. Fleischmann took a silver electrode, roughened the surface electrochemically, and condensed onto it a single layer of pyridine, which is a molecule similar to benzene. Then he bounced a laser beam off the electrode and claimed he could quite clearly observe the pyridine spectrum. In fact, he claimed the technique was a million times more sensitive than anyone had reason to believe.

When he reported the result, in May 1974, many electrochemists refused to believe it. But Fleischmann insisted he was right and continued to lecture on the technique and his results. Eventually his colleagues got around to duplicating his work, and, sure enough, Fleischmann had been right. Triumphs like this led Fleischmann to say that cold fusion wasn't the first time he'd stuck his neck out: "People have often said that I am wrong," he said, "and they have usually had to eat their words."

Curiously enough, Fleischmann himself failed to understand what had brought about his success in Raman spectroscopy. There had been a phenomenal element of serendipity in the experiment. First of all, Fleischmann had intuitively chosen the best possible system for observing the effect—silver and pyridine. If, for instance, he had used a platinum electrode, which is a common choice in electrochemistry, the experiment would have failed. For the next three years, however, Fleischmann seemed to believe that by roughening the silver electrode he had managed to increase its surface area by a factor of a few thousand, and that this increase had generated the millionfold enhancement of the signal.

In retrospect, this explanation sounds implausible. And it was. Richard Van Duyne, a chemist from Northwestern University, did the systematic study of the phenomenon—a study Fleischmann could have done at any point. Van Duyne proved that the enhancement was caused not by the roughening of the electrode but by a resonance created in the electron bonds holding the pyridine molecules to the silver. It was Van Duyne, in that sense, who actually discovered the phenomenon now called surface-enhanced Raman scattering, even though Fleischmann was the first to have observed it.

Nonetheless, it was Fleischmann, much more than Pons, who gave cold fusion its pedigree and credibility. At first glance Pons seemed to be a perfect collaborator for a man with Fleischmann's imagination. Most of Fleischmann's colleagues would listen to his ideas spew forth like sparks from a Roman candle and then say, "That's an interesting idea, Martin; we have to do something about it someday." Pons apparently would hear Fleischmann's ideas and get right to them. He had "absolutely fantastic energy," in the words of David Williams, an electrochemist at the Harwell laboratory who had collaborated frequently with Fleischmann. "You have to respect people with that kind of enthusiasm and energy," Williams said, "because they'll do things that others can't be bothered with doing, things that others would just forget. They make life interesting."

But the Pons-Fleischmann collaboration had a downside. Fleischmann sparked a good number of wrong ideas for every brilliant one. For a healthy collaboration, scientists of such impetuous intellects must be paired with colleagues who are especially discriminating, as well as cautious, qualities that Pons did not possess in great abundance.

Good science is often incompatible with speed, and Pons was both aggressive and fast. At the rate of thirty-six papers a year, mistakes seem unavoidable. After the cold fusion announcement, Fleischmann joked about how the two of them often discussed doing what even he called "insane experiments," which had a lot of truth to it. Barry Miller, a respected electrochemist at AT&T Bell Labs, said they simply had similar personalities. "They're both idea people," he said, "who don't keep reins on each other."

A look at the literature provides some examples. Three Pons and Fleischmann papers, for instance, all published in 1986, claimed remarkable results. Unfortunately, the papers range from dead wrong to recklessly interpreted.

The least explicable of these papers was one in which Pons and Fleischmann claimed to be able to use ultramicroelectrodes to oxidize rare gases, in particular krypton. A remarkable claim, considering that these are also known as noble or inert gases, precisely because they have little or no ability to react chemically. It is extraordinarily difficult to add an electron to an atom of a rare gas, a process called reduction, or to take an electron off such an atom, known as oxidation.

Pons and Fleischmann's claim was profound, and the clue that suggested the error was obvious. It required the same amount of energy to oxidize krypton as it did argon, xenon, and the other noble gases. This makes no sense from an atomic viewpoint. It did suggest that what was actually happening was that Pons was oxidizing a film of water on his electrodes. Although, said one chemist, who preferred to remain anonymous, "It could have been anything." Mark Wightman, a University of North Carolina chemist whom many consider one of the finest electrochemists around, said, "I just don't understand how that paper ever got published [in the *Journal of Physical Chemistry*]. Somebody didn't do their job on peer review." Pons and Fleischmann retreated slightly from the original claim in a later paper, but they never actually admitted that they had been in error.

In another paper, Pons and Fleischmann described an experiment in which they put a voltage across two parallel plates and flowed a solution containing iron beads between the plates. The electric field on the plates, they claimed, induced a field on each bead, and this field then oxidized molecules on one side of the bead and reduced them on the other. This was theoretically thought provoking, but it was contrary to the well-established laws of physics and almost assuredly wrong. For whatever reason, this paper managed to be published twice in the same journal, which also raised questions about the peer review and publication process.

In the most controversial of the three papers, Pons and Fleischmann claimed to have pulled off a feat called gas-phase electrochemistry. This, again, seemed like a physical impossibility.

This discussion requires a brief description of electrochemistry, which, for the sake of simplicity, can be reduced to the study of two physical reactions: those in which chemical reactions generate electricity, as in solar cells, and those in which electrical currents induce chemical reactions, a process known as electrolysis. The fundamental laboratory setup of the discipline has two electrodes, a cathode and an anode, placed in a liquid known as an electrolyte. A voltage is placed across the electrodes, causing a current to flow from the cathode through the electrolyte to the anode. In fact, the molecules in the electrolyte are reduced at the cathode, gaining an electron, and are oxidized, or relieved of an electron, at the anode. For this arrangement to work, the electrolyte must be able to conduct ions, which are molecules with a net positive or negative electric charge. Virtually any liquid into which a salt has been dissolved will serve well as an electrolyte. Gases, however, are unable to support a high enough concentration of ions to conduct electricity, let alone allow conventional electrochemical processes to occur. Any chemist who claimed that he could do electrochemistry in a gas would need ironclad evidence to convince his peers.

Pons and Fleischmann's evidence was considerably less than ironclad. It turned out that their gas-phase electrochemistry only worked when they put a thin film of liquid over the electrodes, which strongly suggested that the liquid was doing the conducting. This was neither particularly new nor noteworthy. Depending on the point of view, the Pons-Fleischmann gas-phase electrochemistry paper was either "bloody neat, but a definite oversell," as Dave Williams of Harwell put it, or "nonsense," as Nathan Lewis, an electrochemist at the California Institute of Technology, called it. Lewis's opinion seemed to be in the majority. Charles Martin, for instance, an electrochemist at Texas A&M University and a friend of both Lewis and Pons, said, "I read this article, and I immediately knew it was wrong."

Pons did this work partially under a research grant for Dow Chemical, which suggests that although Fleischmann's name was on the paper, it may have been Fleischmann's idea and Pons's follow-through. Either way, after what Dow administrators would describe only as a "very unpleasant experience," they cut off support to Pons. As the story is told, once Pons claimed the effect had been established, Dow immediately began filing for patents. When the Dow attorneys found that they lacked sufficient information to do so, they went back to Pons. "That's when the trouble started," said one of the Dow chemists, who requested anonymity. Instead of information for the patent, Dow began receiving letters from C. Gary Triggs, Pons's lawyer, threatening legal action if Dow didn't back off.

Once the company's chemists realized the scientific claims were irreproducible, Dow chose to end the collaboration rather than engage in a nonsensical legal battle. "What's sad about the whole thing," said one Dow administrator, who also requested that his name not be used, "is that Stan Pons is actually a fairly brilliant guy. When he came in to Dow and gave his original presentations, he impressed the hell out of people, and that's one of the reasons we chose to sponsor him." In fact, one of the Dow chemists refereed the gas-phase electrochemistry paper for a journal and suggested it not be published because it was wrong. "Somehow it got through the system and was published virtually as originally submitted," the chemist said.[4]

The Dow case was not an isolated example. In the spring of 1986, a firm called Synthetech Inc., of Boulder, Colorado, hired Pons as a consultant to solve an electrochemical problem. Two months later Pons reported back that he had done so. When the Synthetech chemists tried to replicate his work, however, they could not. Several times they visited the Utah laboratory and, at least once, watched as Pons and his researchers apparently performed the experiment successfully. But again, back in Boulder, the Synthetech chemists failed to make it work. In October, Pons suggested that the problem lay in the Synthetech equipment and sold the company a $6,000 electronic device called a potentiostat, which happened to be of his own design and marketed by his company. With this Stan Pons signature-model apparatus, the Synthetech chemists tried again and failed. Synthetech also hired Carl Koval, an electrochemist at the University of Colorado, and he tried and failed.

By this time, according to Paul Ahrens of Synthetech, the company requested Pons's raw data, which was contractually theirs and which, presumably, would provide enough details to allow the Synthetech chemists finally to reproduce the work. Pons responded with an interim report that, said Ahrens, was remarkable for its incompleteness, repeatedly promising more data but not including it. For six months Synthetech pestered Pons for more details without success.

Finally, in March 1987, Synthetech wrote what Ahrens called "fairly provocative" letters to University of Utah president Chase Peterson and to Jack Simons, chairman of the chemistry department. In return, Synthetech received a threatening letter from Triggs, arguing that the problem was not with his client but with Synthetech's lack of expertise. Eventually, the Synthetech chemists derived their own solution to the electrochemical problem and, for essentially the same reasons as Dow, let the issue with Pons drop.

So Pons and Fleischmann made mistakes, which seemed to be the product of sloppy experimental techniques combined with hubris. Then Pons, working with Fleischmann, appeared to believe that it was better to be rash, and to risk misinterpreting his data, than to lose a chance at fame or fortune. Kevin Ashley, who obtained his Ph.D. under Pons in 1987, said his former adviser seemed to care not so much about whether his interpretation of data was right as about whether he could get his work published. Ashley was not alone, among Pons's former students, in making this suggestion. "Stan was a smart guy," Ashley said, "he did some good science, but he really did some sloppy shit, too, which surprised me. He had established a good reputation, but, like many full professors at big schools, he had a big ego. He always wanted to show other people up. And his big way of showing up other electrochemists was to get more publications than they did."

The competition for funding in academic science is so vicious that scientists with a flair for salesmanship and a slight tendency to overinterpret their results can do well. And Pons did well. Not only were his funding agents his friends, but he was publishing sensational papers. That some of these were wrong, that his scientific claims seemed implausible to begin with, and in fact were downright impossible, was rarely discussed.[5] Chemists who were familiar with these errors refrained from mentioning them to the press even after the cold fusion announcement for the same reasons they had never mentioned them to the funding offices. It is considered in poor taste, to point out that the experimental techniques of a competitor leave something to be desired. That would also be inviting the competitor to return the favor someday. Whatever the reason, these errors hadn't damned either Pons or Fleischmann. In fact, they had barely made their work even slightly less credible. Everybody makes mistakes.

As for the commercial fiascoes, the people at Dow believed that any publicity on the kind of problems they had had with Pons would eventually become bad publicity, which was probably true. And still may be. Paul Ahrens of Synthetech said that they had considered contacting the press after the cold fusion announcement, but they assumed that they were an isolated instance. Just because Pons had given them a hard time did not mean that cold fusion shouldn't be taken seriously. "You never know," Ahrens said.[6]

Alan Bewick, an electrochemist at Southampton who was Pons's thesis adviser and Fleischmann's colleague, recalled that Fleischmann had first aired his theory of cold fusion at Southampton back in 1974 or 1975, "before Stan was even heard of." Although Fleischmann kept talking about it, however, he never got around to trying it, nor did he find a colleague at Southampton who considered it worthy enough to pursue. "When he went to Utah [in the early 1980s]," Bewick said, "it gave him the opportunity to actually try out these mad ideas."

Because Fleischmann was at heart an electrochemist, he had not been constrained by the orthodox thinking of physicists when it came to creating a nuclear fusion reactor. What has made nuclear fusion such a formidable technological challenge is the simple fact that atomic nuclei consist of positively charged protons and neutral neutrons (or, in hydrogen, just a proton) and carry a net positive charge. This repulsive force makes it difficult, if not impossible, for two nuclei to get close enough to fuse, although the repulsion can be overwhelmed if the two nuclei are moving toward each other with sufficient energy. In the interior of the sun and stars, the density of hydrogen and deuterium is so great and the temperature so high—tens of millions of degrees—that fusion reactions occur constantly. (The *Encyclopædia Britannica,* that repository of conventional wisdom, says that in stars, at least, "a minimum temperature for fusion is roughly 10,000,000° kelvin.") In a hydrogen bomb, the temperature necessary to induce nuclear fusion is obtained by first exploding an enclosed fission bomb—an A-bomb.within the H-bomb.

The textbook approach to inducing nuclear fusion in the laboratory— what cold fusion aficionados would refer to as the physicist's approach— has been to take a mixture of hydrogen and deuterium gases, or deuterium and tritium, and heat it in a containment device called a tokamak to 100 million degrees. At this temperature, the gas exists in the form of a plasma (in which all electrons are stripped from the nuclei), and fusion between the nuclei will happen readily. Whether a fusion reactor can ever be made to produce more energy than it consumes depends on whether a dense enough plasma can be heated to a high enough temperature and then contained for a suitable amount of time. It is not easy.

Thinking as an electrochemist, Fleischmann speculated that the modest electric current charging an electrolytic cell would create an enormous concentration of deuterium nuclei inside the palladium electrode.

Perhaps the pressure on the nuclei would be high enough to induce fusion without requiring extraordinary temperatures. Fleischmann believed that strange things happen to deuterium in a palladium lattice, so one never knew.

According to the University of Utah press release (which proved to be a dubious source of information on many points), Fleischmann had been led to attempt this as far back as the 1960s, when he conducted experiments on the separation of hydrogen and deuterium isotopes using palladium electrodes. The idea was that when palladium foils were used as electrodes in cells similar to the fusion cells of twenty years later, the rate at which hydrogen and deuterium were drawn into the lattice of the palladium would differ significantly, an effect known as the electrolytic separation factor. Fleischmann had published one of the seminal papers on this work in 1972. "His interpretation of the data," said the press release, not mentioning what exactly those data were, "indicated it would be worth looking for nuclear fusion reactions."

Meanwhile, Pons, too, had supposedly been puzzled by peculiar data generated by these electrolytic separation experiments. Then, according to the press release, Pons and Fleischmann discussed these findings while driving together through Texas and again on a hike up Millcreek Canyon on the outskirts of Salt Lake City. "Stan and I talk often of doing impossible experiments. We each have a good track record of getting them to work," the release quoted Fleischmann. "The stakes were so high with this one, we decided we had to try it."

> They decided to self-fund the research rather than try to raise funds outside the University because, says Pons, "We thought we wouldn't be able to raise any money since the experiment was so farfetched."
> Working late into the night and on weekends at Pons' University of Utah laboratory, the two improved and tested the procedure throughout a five-and-a-half year period.

This account changed little through the press conference and subsequent interviews. (The only additions were Fleischmann's playful suggestion that a bottle of Jack Daniel's helped the cerebration process and his remark during the press conference that the price tag on their investment was $100,000.)

The operative fact in the release is the propitious "five-and-a-half" years of work. This sets the origin of cold fusion around October 1983, at least one year before the meltdown and some two years before the earliest date given by Steve Jones's competing experiment at Brigham

Young University. It is also a month or so after Pons left the University of Alberta and arrived in Utah and around the time Fleischmann took his early retirement at Southampton and became an honorary research professor. It is, coincidentally or not, the earliest date Utah could establish for cold fusion conception that might avert a legal claim from either Alberta or Southampton.

Although Pons and Fleischmann claimed they had proof of their five years of cold fusion research, they were extraordinarily reserved about demonstrating what it was. Norman Brown, head of the University of Utah Office of Technology Transfer, said he hadn't seen it, but perhaps Chase Peterson or Jim Brophy had. Peterson would not say that he'd seen proof of the five-plus years, only that such a number was needed to establish a patent claim. As for Brophy, although he'd known Pons since he came to Utah in 1983, he first heard of cold fusion in July 1988.[7]

The first time Pons publicly admitted he was working on cold fusion was in September 1987. The University of Utah had arranged for an external review of the chemistry department, a common procedure in academia. Among the chemists brought in were Al Bard, from the University of Texas, one of the premier electrochemists in the country, and Richard Bernstein of UCLA, a veteran of the Manhattan Project who had developed the science of molecular beam scattering. Pons never mentioned fusion to Bard. He apparently did tell Bernstein about it, swearing him to secrecy.

Bernstein was astonished. He had worked on the electrolysis of heavy water with palladium back in the 1950s. How could he have missed cold fusion? He had used the same current densities under slightly different conditions, but they didn't seem that different. He did not believe that he could have missed such an important observation. Now he promised he wouldn't say anything. He couldn't even work on it just to satisfy his curiosity. It would haunt him for the coming year.[8]

On Bernstein's last night in Utah, the chemistry department threw a party. Joseph Taylor, outgoing dean of the College of Science, and Hugo Rossi, his successor, were among the guests. Bernstein told Taylor and Rossi that Pons was doing extraordinary work, but it was necessary to keep it confidential. Bernstein explained that Pons was not yet ready to apply for outside support. Still, if it truly worked, what he was doing would have profound implications and should be supported.

Once Rossi was officially appointed dean, he invited Pons in for a talk. He told Pons about his conversation with Bernstein, and that he was willing to trust Bernstein's judgment. Rossi promised Pons whatever support he could give, financially or otherwise. If Pons needed any help with patents, Rossi added, he should let the university know. Rossi took it on faith that when he was ready Pons would let him know what they were doing. So Pons thanked Rossi and told him that he and Fleischmann were doing just fine.

In April 1988 Jack Simons, chairman of the chemistry department, announced his resignation. Rossi went looking for a new chair and decided on Pons, who had enough seniority as well as support throughout the department. Rossi's one concern was that Pons's new administrative duties would take him away from his research, particularly the mystery project. He therefore insisted that Richard Steiner, the associate chairman, take responsibility for much of the department's paperwork.

Finally, in May 1988, Pons handed Rossi a manila envelope marked CONFIDENTIAL. He and Fleischmann, Rossi was told, were applying to the Office of Naval Research for financial support. Although Pons was already receiving $300,000 yearly from ONR for his laboratory work, he was asking for an additional $100,000 for cold fusion. The envelope contained the research proposal. "Read it sometime when you have the time," Pons said.

Rossi found the proposal, entitled "The Behavior of Electrochemically Compressed Hydrogen and Deuterium," dry and scientific, absent any spectacular pronouncements about what cold fusion would do for the world's ills. To Rossi, who was a mathematician unschooled in either nuclear theory or experimental science, Pons seemed to be pursuing a plausible but improbable project.

The larger implications of the project didn't dawn on Rossi until the middle of the night. "All of a sudden," Rossi recalled, "I woke up and I said, 'These guys are building a hydrogen bomb in the basement of the chemistry building.' "

He called Pons early the next morning, knowing the chemist would be in his lab, taking advantage of the hour to work undisturbed. "What are you doing down there?" Rossi asked. "Are you building a bomb?"

"Well," said Pons, "I suppose you could put it that way."

"Is there any danger?"

"No," said Pons. "Whatever there is, it seems it comes across at a low enough level. Of course, you're never sure, but we feel quite comfortable."

"Maybe you should move it off campus," Rossi suggested.

"Don't worry about it," Pons said. "It won't be necessary."

CHAPTER 2

THE COMPETITION

*O*nce Stan Pons submitted his proposal to his benefactors at the Office of Naval Research, the cold fusion affair took on an aura of inevitability. Although ONR had a reputation for funding speculative research projects, either Pons or Robert Nowak, his funding officer, decided that ONR was not the right place for cold fusion. Pons told Hugo Rossi that he feared the Department of Defense, of which ONR was a part, might classify cold fusion. They might realize the potential military uses of an invention that, if it wasn't a hydrogen bomb itself, would still produce tritium, a necessary and valuable component of hydrogen bombs.

Since Pons preferred to think of cold fusion as an energy source, not a weapon, he sent the proposal to his friend Jerry Smith, a funding agent in the Department of Energy's physical chemistry program. At one time Smith had been program manager for Pons at ONR, and his name appears as a coauthor on several of Pons's journal articles. Smith, too, felt that cold fusion was inappropriate for his office and suggested that Pons submit his proposal to the Office of Advanced Energy Projects at DOE, run by an administrator named Ryszard Gajewski (pronounced Richard Guy-EV-ski).

In September 1988, Pons told Rossi that the proposal had just been sent to Advanced Energy Projects, and it might be five or six months before he got his answer.[1] It would take less time than that, however, for Pons and Fleischmann quickly and irrevocably to lose control over the discovery of salvation.

"Infants and orphans" was the way Gajewski liked to refer to the research he supported at the Office of Advanced Energy Projects (AEP). To certain scientists, it was also known as the Office of Off-the-Wall Ideas. The office had been inaugurated in 1977 to do for the Department of Energy what the Office of Naval Research and the Defense Advanced Research Projects Agency did for the Pentagon. The Office of Advanced Energy Projects would support the kind of speculative ideas that might not pass peer review in the conventional programs but if given a glimmer of a chance might pan out—in other words, long shots. In particular, AEP supported exploratory research on power and energy projects that didn't fit into any of the other pigeonholes of the DOE's Basic Energy Sciences Division. The division had a yearly budget of $10 million and a professional staff of one—Ryszard Gajewski.

Gajewski was a diplomatic, old-world physicist from Warsaw, who had emigrated to America with his family in 1969. He had settled in Boston, spent two years at MIT, another year at Brandeis, and then a handful working for a technology company called American Science and Engineering in Cambridge. In 1977 he moved to the suburbs of Washington to become the first director of AEP.

Gajewski had an unusual disposition for the DOE. Funding officers, after all, are civil servants. They are not famous for their creativity, and they can be short on vision and imagination. It could be argued that had these funding agents been productive scientists, they would have stayed in research. Gajewski considered himself, and in fact seemed to be, a man of vision. He ran AEP less like a banker and more like a Medici. He took personal satisfaction in shepherding his projects to success and spoke of them as though they were his children—thus, infants and orphans.

Since 1977 Gajewski had been supporting eight to twelve new projects a year, for up to three years each. At any one time he might have thirty to forty projects under his purview. These ranged from novel storage batteries to sophisticated oil extraction processes and very high-tech laser beams. If, after three years, his orphans could not entice one of the other DOE funding offices to provide for them, it was a good bet that they never would. Three years and out was the program's bottom line.

During his tenure Gajewski had managed to support two projects for six years or longer. The first was an X-ray laser project for hot fusion, which was being pursued by a Princeton physicist.[2] The second was known as muon-catalyzed fusion, the forerunner of what has since been called cold fusion. Muon-catalyzed fusion was one of Gajewski's visions. He equated it to his own background. "Like an immigrant's child," Gajewski once wrote of muon-catalyzed fusion, "in order to succeed in a big way we must prove that we have something clearly superior to offer, something the society simply cannot ignore."

The point man for this program was Steven Jones of Brigham Young University. After the March 23 announcement, when the press discovered that Jones had also been working on cold fusion and were trying to establish the nature of the relationship between Pons, Fleischmann, and Jones, Gajewski would tell reporters that if he had to choose three people to trust in the world, Jones would be one of the three. Left unsaid, of course, was that neither of the other two would be Pons or Fleischmann.

Jones's colleagues at Brigham Young were of a similar opinion. Jae Ballif, the provost, considered Jones a man so open, so honest and without guile that it was simply impossible not to believe in him—unless, of course, you had guile in your own heart. And Lamond Tullis, who was then associate academic vice president, said that Jones had few peers with respect to integrity and honesty, that he was constitutionally incapable of duplicity. Jones had been at BYU since 1985. He attended devotional classes regularly and kept notes on the classes in his logbook. He never worked on Sunday, devoting that day to his wife and six children and the Mormon church. Even by the high standards of BYU, Steve Jones appeared to be devout.

To Pons, Fleischmann, and various of the administrators of the University of Utah, by contrast, Jones came across as (literally) too good to be true. He was the man who doth protest too much. Steve Jones would actually talk about following the Golden Rule. Or he'd say, "Eventually we're going to die or we're going to end up someplace, and who discovered what, when, is probably not going to be very important. What you do to advance truth will be." He insisted that Pons and Fleischmann simply never understood the depths of his Christian beliefs. Joe Taylor later described Jones as a "Boy Scout," a description Jones may not have considered uncomplimentary.

It was Jones's tenure at BYU—and the endless competition between

BYU and Utah—that lifted cold fusion from a scientific controversy to an emotional one. And it certainly contributed to the evolution of the cold fusion affair that BYU and the University of Utah had been rivals well before cold fusion came along. These two schools are Utah's equivalent of Harvard and Yale, or Oxford and Cambridge. The locals referred to BYU as the Y and the University of Utah as the U. Whereas the Y was almost exclusively Mormon, owned and run by the Church of Jesus Christ of Latter-day Saints, the U was only half Mormon, was run by the state, and aspired to be a cosmopolitan center of learning. And the U was in Salt Lake City; the Y was forty-five miles south in Provo. This proximity seems almost perverse: as Peter Dehlinger, the attorney who filed the first patent applications for the University of Utah, put it, "With an entire universe, you'd think the gods would've put them a little further apart."

Jones, who would turn forty two days after the Utah announcement, had graduated magna cum laude from BYU. He was justifiably proud of his academic record. (On his curriculum vitae, circa 1989, when he was already an associate professor of a major university, he mentioned not only the high school he attended, Bellevue H.S. in Bellevue, Washington, but his grade point average, 4.0.) He spent twenty-six months in France and Belgium on the religious mission encouraged of young Mormon men, then pursued his graduate studies in physics at Vanderbilt University. Bob Panvini, who was Jones's thesis adviser, observed that what made Jones special was not so much that he was a brilliant student but that he was remarkably efficient. He managed to fulfill his academic obligations while spending weeknights and Saturdays and Sundays with his family. Most successful physicists, especially experimentalists, tend to be workaholics who can't imagine anything coming before physics— certainly not their family or their God.

Jones did his thesis research at the Stanford Linear Accelerator Center, searching for hypothetical particles that almost assuredly do not exist and would likely not have been found if they did. As far as Jones was concerned, this experiment taught him that physics is as much a competition with his fellow man as it is a profound search for truth, a revelation which deeply troubled him. One of the senior members of his experimental team worked out a preliminary analysis and convinced himself, for long enough to regret it, that they had discovered something. Apparently news of the erroneous discovery made it into several major newspapers. Jones was assigned to do the thorough analysis and, of course, proved that they had found nothing. The experience soured him.

"Look," he would say to his friends. "I'm real disillusioned here. All we're after is the facts, and yet there's so much animosity generated over this, and ego."

"Well," his friends argued, "ego is a large part of it." Jones agreed.

After getting his doctorate, Jones bounced around for a few years trying to find a home in physics. He spent five months at Cornell, where he was once again "disillusioned by all the ego tripping." He spent a year or so at the Los Alamos National Laboratory doing intermediate- to low-energy physics. Finally, in 1979, he went to work for the Idaho National Engineering Laboratory, INEL, as a senior engineering specialist. This career path would not be described by many physicists as upward mobility.

In 1979, however, the INEL physicists were looking for a source of neutrons to study the kind of radiation damage that would plague fusion reactors of the future. Jones recalled that as a graduate student he had read about muon-catalyzed fusion in a volume entitled *Adventures in Physics*. The phenomenon had been discovered serendipitously in 1956 by Luis Alvarez and his collaborators at the University of California at Berkeley. These physicists had built one of the world's first hydrogen bubble chambers, which was used for studying the collisions of subatomic particles and for which Alvarez would later win the Nobel Prize. In the liquid hydrogen of the bubble chamber, they detected, with some astonishment, the signs of nuclear fusion.[3]

What made this so remarkable was that the liquid hydrogen existed at a temperature of $-250°$ C, whereas the temperature at which fusion occurs in the core of the sun runs to tens of millions of degrees. As Alvarez was to learn—with some guidance from, among others, Edward Teller—the fusion had been sparked by ephemeral subatomic particles known as muons. These particles are identical to electrons except that they are, for whatever reason, 207 times more massive. Because of the similarity between the two, the muon can replace the electron in a hydrogen-deuterium molecule. With its extra mass, the muon can, in effect, shield the repulsive force between the hydrogen and the deuterium long enough for the two to fuse into helium.

When Steve Jones read about Alvarez's research, he thought that if he could induce fusion through this muon-catalysis process, he would get neutrons, which might satisfy the needs of his employers. And maybe if he could get fusion he would also get unlimited energy. Jones seemed to believe that anything was possible.

Alvarez had followed a similar, albeit shorter, train of thought, back in 1956. As he recalled in his memoirs thirty years later:

For a few exhilarating hours, we thought we had solved mankind's energy problems forever. Our first, hasty calculations indicated that in liquid hydrogen-deuterium a single negative muon should catalyze enough fusion reactions before it decayed to supply enough energy to operate the accelerator to make more muons, to extract the necessary hydrogen and deuterium from the sea, and to feed the power grid. While everyone else had been trying to control thermonuclear fusion by heating hydrogen plasmas to millions of degrees, we seemed to have stumbled upon a better than break-even reaction that operated at minus 250 degrees Celsius.

Alvarez later did a more complete calculation and discovered that each muon would have to catalyze over 1000 fusions to generate enough energy to break even. He never observed more than two fusions from any of his muons, and that only occurred twice. He let the subject drop.

Over the years, an occasional physicist would follow in his footsteps, take a new look at muon-catalyzed fusion, realize its promise, work out the calculations, and abandon it. Two years before Jones came along, a pair of Soviet theorists, Leonid Ponamarev and S. S. Gerschstein, predicted that with the right mixture of deuterium and tritium, rather than hydrogen and deuterium, at the right temperature, fusion could be induced to occur thousands of times more rapidly than previously thought. It wasn't unreasonable to think that if the Soviets were right muon-catalyzed fusion could be a commercial energy source. It was a long shot, which was enough for Steve Jones.

With support from EG&G Idaho, the contractor that ran INEL, Jones set up an experiment at Los Alamos, using the Los Alamos Meson Physics Facility to provide his muons. His contribution to the field would be to try a fifty-fifty mixture of deuterium and tritium and see whether the Soviet theorists knew what they were talking about.

Jones also took muon-catalyzed fusion to the DOE and reached Ryszard Gajewski, in much the same way Pons and Fleischmann would six years later. As Gajewski remembered it, "He had nowhere to go. He had tried the magnetic fusion program, and they had laughed him out of court, and he tried inertial fusion, and they laughed him out of court, and then he was referred to me, and this was just a perfect match. It was a new idea, and a relatively simple experiment could prove it or disprove it." Beginning in March 1982, Gajewski signed Jones up for three years and a shade under $1,300,000.

Jones observed his first encouraging results during the long Thanksgiving weekend of 1982, when his experiment, to everyone's apparent surprise, confirmed one of the Soviet predictions. By varying the temperature of the gas, they could increase or decrease the number of fusions catalyzed by each muon. In 1983 Jones and his colleagues published an article in *Physical Review Letters* reporting that they had achieved the remarkable number of 100 fusions per muon. At least once, they said, they had counted as many as 150.[4]

From then on Jones believed in the promise of muon-catalyzed fusion. The research would make his career and his name in physics, but progress would be slow at best. By 1985, after his first three-year stint with Gajewski had come to an end, Jones's fusion research was teetering on the brink of extinction. Although Gajewski managed in September 1985 to get Jones two more years of support, at a quarter of a million dollars a year, no one knew how long he could keep the program going.

On March 12, 1986, Jones gave a seminar to his BYU colleagues on his fusion research. One member of the audience was Paul Palmer, a journeyman physicist in his late fifties. Palmer's kaleidoscopic career included aerospace and mechanical engineering, ballistics, meteorology, atmospheric physics, and, for the last fifteen years, acoustical physics—with a little geology on the side. "I pick up pretty rocks when I see them," he said. To his résumé he was about to add cold nuclear fusion.

Over the years Palmer had concluded that much of the geology taught in textbooks is "a bunch of baloney." A case in point was the conventional wisdom about the source of the internal heating of the earth. These geology books attributed the heat to radioactivity from long-lived elements such as uranium, but Palmer said he didn't buy it. After all, volcanism is driven by this heat source, but neither volcanic ash nor lava is radioactive. Volcanic ash and lava do, however, seem to have an overabundance of helium 3, the nucleus of which consists of two protons and only one neutron. Garden variety helium, or helium 4, has two protons and two neutrons.

Palmer reasoned that since helium is literally lighter than air, most of it should have escaped into space in the early history of the planet. Whereas helium 4 is created naturally in the earth in the radioactive decay of heavy elements such as uranium, helium 3 is created by nuclear fusion. This led Palmer to the conclusion that a planet without naturally occurring fusion should have very little, if any, helium 3. So he found it curious that time and time again geophysicists seemed to find inexplicably high ratios of helium 3 to helium 4 in various metals.

Thus far it sounds both simple and reasonable, but Palmer's theorizing revealed him as an amateur in this field. Professional geophysicists suggest that the earth accumulated a fair share of helium 3 during its formation, which is why helium 3 to helium 4 ratios identical to those on earth will be found in some meteorites. And helium 3 can also be formed from cosmic ray bombardment. If you come across a metal with a high ratio of helium 3 to helium 4, it's a good sign that the metal has been lying around on the surface of the earth for a while, where it is fair game for cosmic rays. (It would have been an even better target if it had been lying around the top of a volcano.)

Some metals have high helium 3 ratios because of the absorption of radioactive tritium from atmospheric bomb tests. For instance, water from the atmosphere will catalytically reduce on the surface of iron. The iron will oxidize—which is to say, rust—and the hydrogen will then diffuse into the iron. If there's tritium in the water, then the tritium will also diffuse into the iron and eventually decay to helium 3. This was a published phenomenon, but it hadn't disseminated into college textbooks, so many generalists—such as Palmer, apparently—were not aware of this work.

To Palmer, muon-catalyzed fusion was a revelation. If fusion could be catalyzed by muons, he figured, it could also be catalyzed at very low levels by other factors. Pressure? Who knew? One of Palmer's notebook entries refers to it as "fusion catalyzed by something or other." Palmer's reasoning went like this: deuterium exists in abundance in seawater, and an abundance of seawater is carried down through what are called subduction zones deep into the center of the earth. If "something or other" catalyzed fusion deep in the bowels of the earth, that would explain the heat of volcanism and, simultaneously, the excess helium 3.

Palmer related this logic to Jones, who found it appealing. Jones had just published a paper with Clinton Van Siclen of the Idaho National Engineering Laboratory on piezonuclear fusion—Jones's term for fusion induced by extraordinary pressure rather than extraordinary high temperatures.[5] *Piezo* is the Greek root for pressure. Jones and Van Siclen had calculated the fusion rate of deuterium at room temperature and atmospheric pressure. They concluded, as had the others before them, that it would be, for all intents and purposes, zero. Jones and Van Siclen had also speculated briefly on fusion at high pressures and low temperature, and whether this might be the source of heat in Jupiter, where core pressures would reach 60 million atmospheres. Again, unfortunately, the fusion rate appeared to be "many orders of magnitude too small to be a significant source of energy." Still, Palmer suggested that their calculations never took into account the presence of "something or other" as a catalyst.

A few weeks after Palmer broached his theory to Jones, they came upon a paper by Boris Mamyrin, a Soviet researcher, who found excessive amounts of helium 3 in nickel foils. Fusion? Why not? In a memo dated April 1, 1986, Jones wrote, "Could it be that metal hydrides provide an environment conducive to confinement and fusion of hydrogen isotopes?"

On April 7, Jones met at BYU with Palmer, Bart Czirr, the resident radiation detection expert, and Johann Rafelski, a theorist who was now collaborating with Jones on the muon-catalyzed fusion work. The four scientists discussed various strategies for catalyzing fusion at room temperature. Later Jones liked to call this meeting "the brainstorming session." The scientists discussed using diamond anvil presses to condense deuterium, or even electric charges or lasers to shock deuterium atoms into fusing.

Jones's notes for the day, as was his style, were cryptic. His handwriting bordered on the illegible. And, if he was then planning to use electrolysis to condense deuterium in a metal and induce fusion, as he would claim later, he never actually wrote down the word *electrolysis*. What is indisputable is that he scribbled a list of elements: "Al, Cu, Ni, Pt, Pd, Li . . ." And next to Pd, palladium, and Pt, platinum, were the portentous words "dissolves much hydrogen." And Jones did, at Rafelski's suggestion, take the lab book to the BYU patent attorney, Lee Phillips, and ask that the page be notarized.

Three years later, and several weeks after the March 23 announcement of the discovery of cold fusion, the BYU press office released an official history of "piezonuclear" fusion, which was now simply Jones's term for cold fusion. This documented the progress of the BYU cold fusion research program, with the aim of dispelling Pons and Fleischmann's accusations that Jones had somehow pirated the idea from them. The account described this April 7 meeting as the beginning of "Brigham Young University's experimental program." This made the BYU effort sound like a concerted three-year program, which is how Jones described it later to Pons and Fleischmann, and later still to reporters. Such was not the case.

On May 13, 1986, Jones submitted his annual progress report on muon-catalyzed fusion to Gajewski and included notes on the piezonuclear fusion ideas. Gajewski, in return, gave Jones the go-ahead to spend a share of his muon-catalyzed fusion money on piezonuclear fusion experiments.

As Jones continued working on muon-catalyzed fusion, Paul Palmer went to work on piezonuclear fusion. He recruited Rod Price, one of his graduate students. They began with a search of the literature, finding, said Palmer, "a jillion books on hydrogen in metals. . . . And so all of that stuff we collected together and decided, wow, it's easy—how do you get hydrogen in metal? You just use electrolysis."

So Palmer and Price, without any training in electrochemistry, began electrolysis experiments. They built the electrochemical cells, as Price put it, "the way you do it in junior high when you want to make hydrogen and oxygen." They used electrodes of copper and nickel because those were the metals they had readily available. An equal mixture of light water and heavy water, with a pinch of sulfuric acid, served as the electrolyte.

In his record of the experiment on May 22, Palmer seems to have been fascinated by the many colors of electrochemistry:

> We constructed an electrolytic cell in a test tube to try to get hydrogen into metals. . . . After 8–10 hours, a heavy green coating built up on cathode ($-$). This flaked off when dry. The cathode was nickel plated underneath in a spotty grey-black silver coating. There was green gelatinous stuff in liquid.

Using Bart Czirr's detector, they looked for gamma rays, which could signify piezonuclear fusion. If the deuterium and hydrogen were really fusing, one end product of this reaction would be a helium 3 nucleus and a neutron, and the neutron would then be captured in the electrolyte and emit a gamma ray. "The results," Palmer wrote, "were completely inconclusive."

On May 27 they went at it again. "At least the trend is in the right direction on paper," wrote Palmer, "even though it is not statistically significant."

Toward the end of June, Palmer and Price jury-rigged an apparatus to vacuum-load deuterium into a nickel foil. This was even easier than electrolysis: heat the nickel in a vacuum chamber, flood the chamber with deuterium gas, then cool the chamber down, during which time the nickel soaks up the deuterium. Then remove the deuterated nickel and place it in front of the gamma detector. It registered no gamma ray signal at all.

Finally, on July 7, an electrolysis cell appeared to emit a gamma ray signal. Palmer later insisted that this signal was as clear and distinct as that which Jones insisted to Pons and Fleischmann he was ready to publish in the winter of 1989. Palmer commented that after Pons and Fleischmann held their press conference and Jones then published his own results, "We said, 'Well, we should have published [our data] back in 1986 and we would have avoided all this flap.' "

This may have been an unfortunate commentary on the value of the 1989 data, because Czirr, who had been building and using detectors for thirty years, said the 1986 data were "nothing worth mentioning" and contained "no real hint of anything."

This was Czirr's business. Palmer considered Czirr the "dean of neutron measurement specialists in the world," a slight exaggeration. Still, Czirr had been in the field a long time. He had spent a good portion of that time at Lawrence Livermore National Laboratory working with neutron counters in the weapons division and for the last twelve years had been a consultant of a sort for BYU. He built particle detectors for a living and sold them to any interested parties, which he described as a nerve-racking livelihood for a man with a dozen children. "I've got a lot of freedom to fail," he said, "and a lot of freedom to starve to death. We just hustle. Keep one step ahead of the wolf."

Over the years Czirr had developed a healthy respect for how hard it is to do good physics. "I've been burned a lot of times on seeing strange things that look wonderful," he said, in his pithy way of putting science in perspective. "You can get fooled real easy. This is a very subtle field. You're looking for small effects in a big sea of background. Of course, that's true of almost any frontline research. You're looking for small effects. If they were big effects, somebody would have seen them long ago."

By the end of September 1986, Palmer and Price had run a few more cells with similar results. Palmer believed they were encouraging. Czirr believed they were not. And for the time being the experiments were abandoned.[6]

While Palmer pursued his cold fusion studies, Jones continued working on muon-catalyzed fusion, which had become a collaborative effort with Johann Rafelski acting as the theorist. The two had been working together since shortly after the Thanksgiving Day breakthrough, and they made a curious partnership.

Jones first became interested in Rafelski after the theorist suggested, like Ponamarev and Gerschstein before him, that tritium and deuterium would be the "name of the game" in muon-catalyzed fusion. After proving that this was indeed the case, Jones sent a note describing his experiment and its result to Rafelski, who then was working at the

University of Cape Town in South Africa. By the summer of 1983, Jones and Rafelski met and began their collaboration.[7]

Rafelski had led a peripatetic life. He was born in 1950 in Poland, where his Jewish parents had survived the Holocaust. The family emigrated to Germany in 1964, where Rafelski was educated; he received his Ph.D. from the University of Frankfurt. From there, Rafelski worked at the University of Pennsylvania, at Argonne Labs in Illinois, at the European Center for Nuclear Research outside Geneva, then went back to Frankfurt, then to Cape Town, before he took a tenured position at the University of Arizona in 1987.

This travelogue appears to be at least partly a result of Rafelski's personality and style of doing physics. Rafelski was often described, even by his sympathetic supporters, as irresponsible and unprincipled, which made his collaboration with the pious Steve Jones all that much harder to understand.

Some physicists considered Rafelski an outright charlatan, while others graciously observed that his personality was his own worst enemy.[8] Berndt Muller, a Duke University theorist who had known Rafelski for twenty years, said Rafelski held many of his colleagues in contempt. "He probably has respect for less than five percent of scientists," said Muller, "which is a very small fraction. And he shows the others that he has no respect for them. I know people who say they would accept his behavior if he were an exceptional scientist, but he is not."

What seemed to anger physicists the most about Rafelski was his impudence, or gall, depending on how kindly one took to it. Rafelski flittered around physics like the quintessential gadfly. He would suddenly appear with a spanking new solution to an intractable problem, or an unconventional theory to explain away bothersome experimental data. Then he would shrug his shoulders and move on when someone took the trouble to prove him wrong. Ponamarev, the renowned Soviet theorist, described Rafelski's theoretical style as "white noise," a kind of meaningless drone like the static on a radio.[9]

Ryszard Gajewski, curiously enough, was perhaps Rafelski's most avid supporter. "Rafelski is world class," Gajewski said. "He has many enemies, because he knows he's good and doesn't hide it. He is very creative, oftentimes wrong, but he is the first to admit it." It seemed to be common knowledge among the muon-catalyzed fusion crowd that Rafelski served as Gajewski's most trusted adviser and that Gajewski would not fund any theory proposals, and maybe even most experimental proposals, without Rafelski's stamp of approval.

One theorist in the field described Rafelski's relationship with Gajewski as "absolutely frightening, almost demonic." Muller, who had collaborated with Rafelski over the years, said he believed the two men's closeness stemmed partly from Gajewski's sense that Rafelski was a fellow visionary. Then there was the Polish connection. Both Rafelski and Gajewski liked to recall their first encounter, when Gajewski discovered to his delight that Rafelski was also a Polish émigré who still spoke fluent Polish.

In 1987 Jones and Rafelski wrote up muon-catalyzed fusion for *Scientific American,* and the article established the two physicists as the gurus of their field. *Scientific American* happens to be one of several lay publications—*The New York Times* is another—that bestows a certain stamp of legitimacy in science. In their article, entitled "Cold Nuclear Fusion," the physicists pitched muon-catalyzed fusion as though only minor problems needed to be overcome before its future in the energy business was assured: "The electron-like particles called muons can catalyze nuclear fusion reactions, eliminating the need for powerful lasers or high-temperature plasmas. The process may one day become a commercial energy source."

The reality, however, was that the muon-catalyzed fusion program was going nowhere. Jones had made little progress after Thanksgiving Day, 1982. And a competing collaboration at the Paul Scherrer Institute (PSI) in Zurich demonstrated that the basic physics of muon-catalyzed fusion condemned it to being little more than a physical curiosity. The *Scientific American* article represented not only mind over matter but also revisionist history.

Physicists familiar with both experiments observed that the PSI experiments, which were initiated a year later than Jones's, were much more professional. More important, the PSI conclusions were much better supported by their data. Even Jones's collaborators said that Jones and Rafelski tended to come to conclusions that seemed not just premature but unwarranted by their own data. In effect, wishful thinking.

One crucial measurement characterized this wishful thinking. It was of a variable known as the sticking probability—the odds that a muon would get caught up in the helium atom after fusion and be unable to catalyze any more fusions. If this probability was high, the muons would never catalyze enough fusions to make muon-catalyzed fusion worth the money. If it was low, this technology might have a future.

The PSI group reported that the sticking probability was not that low and agreed reasonably well with theory. Jones's data actually looked very much like the PSI data, if measurement errors were included, but the trend of the results were very different. The PSI data showed only a slight density dependence, in agreement with theory, while Jones's data points dropped significantly as the density of the deuterium-tritium gas increased. Thus, if they could increase the density of the gas further, anything could happen. David Jackson, a UC Berkeley theorist who wrote a seminal paper on muon-catalyzed fusion in 1959, and who served as the field's elder statesman, observed that Jones seemed to be of two minds on the subject: when Jones presented his data at meetings attended by the PSI competition, he would say that his data were similar to theirs. But, "when you hear Steve Jones talk about it without these other people around," said Jackson, "then he's trying to sell you that the sticking probability is dropping like a bomb, and commercial fusion is just around the corner."[10]

None of Jones's collaborators, however, seemed to believe that this hyping of muon-catalyzed fusion was the result of any dishonesty. Al Anderson, who did much of the data analysis on the experiments, suggested that Jones may have been constitutionally incapable of skepticism. "You know what W. C. Fields says," said Anderson. " 'You can't fool an honest man.' But I have come to the conclusion that sometimes an honest man can fool himself." Anderson, who had worked with Jones for a decade, observed that religion may also have played a role. It is rare for scientists to be deeply religious; to be so requires taking certain aspects of the universe on faith, which is in contravention to the working philosophy of science. "You have a person who has a special relationship with God," Anderson suggested, "and sees God leading him to a discovery. And when God is leading you to a discovery, it has to be there."

Arguably Jones and Rafelski were in an untenable position, which had as much to do with finances as with science. They could only procure funding in America from Gajewski, and he could only provide it so long as commercial fusion appeared to be close to the next corner. That was the institutional condition of his support.

With the exception of Jones and Rafelski, virtually everyone working in muon-catalyzed fusion agreed that the reaction was at least three to five times too rare ever to be a viable energy source on its own. In other words, if a muon-catalyzed fusion reactor could be built, and all the engineering problems could be worked out to perfection, the reactor would still produce less then half as much energy as would be needed to run it. This is still a near miss, by Luis Alvarez's definition, but this miss exists in the essence of the quantum mechanics that control this subatomic world. It cannot be bridged by efficient engineering or technological breakthroughs. Said Dave Jackson, "It's one of those tantalizing things where the basic science dooms you even though you're only a factor of two or three away from pay dirt. That's something engineers have a hard time believing. They're used to missing by a factor of one hundred, but then you nibble at it for ten years and finally make it. But these things aren't nibble-able because they're happening down at the depths of the fundamental process."

As the *Scientific American* article demonstrated, however, Jones and Rafelski still had nearly unwavering faith in muon-catalyzed fusion. At least, they appeared to. In September 1987, two months after the *Scientific American* article appeared, Gajewski managed to get Jones two more years of support at $208,000 per year. And in March 1988, six months after Rafelski began at the University of Arizona,[11] Gajewski awarded him over $975,000 for three years to do "energy related applications of particle theory." It was an extraordinarily large grant for a theorist.

It was about this time, however, that Gajewski was prodded by a DOE advisory panel to adhere strictly to his self-imposed funding limit on the muon-catalyzed fusion work. Three years and out. By now Gajewski had committed the DOE to supporting muon-catalyzed fusion for seven years and almost $3 million (if one included Rafelski's three-year grant). With all this money, the technology was little closer to becoming a commercial energy source than it had been on Thanksgiving Day, 1982.

In April 1988, Gajewski informed Jones and Rafelski of his bind. Without dramatic improvement in the technology and clear indication of its viability as an energy source, he'd have to cut them off. As a last recourse, Gajewski requested an outside review by a group of physicists known in government circles as JASON.

JASON is an elite, quasi-secretive brain trust that serves as an independent advisory group to the government. Composed of the best and brightest physicists in the country, with a sprinkling of mathematicians and representatives of other disciplines thrown in for exigencies, it has all the arrogant selectivity of a prestigious private club. The membership includes at least half a dozen Nobel Prize winners. And no one asks to join JASON. JASON does the asking.

For six weeks every summer, JASON members meet in La Jolla, California, to ponder technical questions for the government. The host organization is the Defense Advanced Research Projects Agency, but JASON will work by contract on problems, classified or nonclassified, for a variety of agencies, from the Federal Aviation Administration to the Department of Energy. Muon-catalyzed fusion was tailor-made for JASON, and Gajewski had it put on the 1988 summer agenda. If JASON said the technology had promise, he might be able to save it.

JASON met at the end of June. Steven Koonin, a nuclear physicist from Caltech, led the eight-physicist panel that had been selected to do the muon-catalyzed fusion review.[12] Steve Jones flew out to La Jolla and spent two lengthy sessions with Koonin's experts, discussing muon-catalyzed fusion and answering questions. Jan Rafelski also appeared before the group, as did one of the leading physicists from the PSI experiment.

On August 1, Gajewski called Jones and said that he had heard from one of the panel members, Doug Eardley, a theoretical astrophysicist from UC Santa Barbara. Eardley told Gajewski that muon-catalyzed fusion provided "no viable path to energy."[13] Over the next few weeks, Jones took to reconsidering his future. He considered submitting a research proposal on piezonuclear fusion to Gajewski, and he mentioned it to Rafelski.

His group had done virtually nothing on cold fusion in the intervening two years, but Jones hadn't forgotten it. In the spring of 1988, two of his undergraduates, Paul Dahl and Paul Banks, had written term papers on the subject. (Both papers were reconstitutions of Palmer's geophysical arguments.) And on March 25, 1988, Palmer had sent his now two-year-old deuterated metal samples to Harmon Craig at the University of California at San Diego to look for helium 3 and tritium. Craig was a geophysicist and a professional aficionado of helium isotopes. He never got around to testing the samples. "He had other fish to fry," said Palmer. Craig kept them for a while before foisting them off on Al Nier, an expert in mass spectroscopy at the University of Minnesota. Nier analyzed them in the summer of 1989, as Palmer recalled, and he did not find anything.

Palmer, more than anyone, couldn't put piezonuclear fusion behind him. Through the summer of 1988, he had occasionally explored alternative ideas for catalyzing fusion. He seemed willing to entertain anything that might result in a few neutrons.

On July 6, 1988, Palmer noted in his lab book a method to produce fusion by loading a fine metal wire with deuterium gas and then somehow exploding the wire. This was actually a variant on hot fusion because it required creating a plasma to initiate the reaction. Another Palmer idea was to pass a huge current across the deuterated wire. Perhaps fusion would be induced by a shock wave in the wire. Palmer was frustrated, as his lab book notes indicate: "I don't know the experiment to demonstrate catalyzed fusion. That is the story of my life—a series of great ideas and a series of fizzles. I never pursue, complete, and publish. As a result I never contribute. It is the finisher who contributes, not the dabbler. Well—so! What now? Nothing is new."

Palmer occasionally told his students that the piezonuclear fusion program had been suspended, while Bart Czirr was developing a neutron detector that might be sensitive enough to observe the infinitesimal neutron radiation that a cold fusion cell might emit. Jones then used the detector to explain why his cold fusion research had effectively shut down after September 1986. "Things really slowed down here dramatically," Jones told reporters. "At that time, we had decided there was something interesting, but we had to develop the neutron spectrometer before we [could] make progress scientifically."

Unfortunately, Czirr saw it differently. He even recalled that in 1986 he had submitted a proposal to the Department of Energy asking for $500,000 to complete work on a neutron detector. He was turned down. From time to time, Czirr would tell Jones that he was building a "whiz-bang detector" that could be useful for muon-catalyzed fusion. Jones never took him up on it. Czirr even sent a proposal to Gajewski asking for support to build his detector, but Gajewski said it wasn't appropriate for his program. "Irony," said Czirr laconically, "knee-deep." Occasionally Czirr attended Jones's fusion group meetings, hoping something would come up that would need a detector. This, he said later, was not because he believed in cold fusion but because he was looking for funding anywhere he could get it.

In any case, before Jones could compose his own proposal to study piezonuclear fusion, Gajewski sent both him and Rafelski a proposal to review entitled "The Behavior of Electrochemically Compressed Hydrogen and Deuterium," by B. Stanley Pons and Martin Fleischmann. Sometime after Jones had received it, Paul Palmer had the following discussion with him:

"I've got this strange proposal that disturbs me," Jones said. "I don't know what to do about it."

"Oh, what's it about?" Palmer asked.

"Well, it's confidential," said Jones. "I better not tell you. But it could be a possible conflict of interest."

"Well," said Palmer, "who's it from?"

"I wish I hadn't told you," said Jones. "I wish I hadn't said anything about it."

"Well," Palmer said, "I now know what it's about or you wouldn't talk to me about it. It's about cold fusion."

End of conversation. Jones wouldn't say any more.

CHAPTER 3

AUTUMN

1988

Scientists try to exonerate themselves in advance from possible charges of filching by going to great lengths to establish their priority of discovery. Often this kind of anticipatory defense produces the very result it was designed to avoid by inviting others to show that prior announcement or publication need not mean there was no plagiarism.

ROBERT MERTON,
The Sociology of Science

S hortly after March 23, 1989, the BYU public relations office distributed an official history of piezonuclear fusion research at BYU. Its purpose was to protect Steve Jones from any possible allegations of conflict of interest or worse—scientific piracy.

This account, which was compiled predominantly by Jones, cited a fusion group meeting on August 24, 1988, during which Jones and his colleagues discussed their piezonuclear fusion program. (This was approximately one month before Jones received the Pons-Fleischmann proposal.) The account asserts that from August 24 onward the fusion group's program was "vigorously" pursued. Jones told reporters, "From that day [August 24] we were essentially 100 percent working on this other piezonuclear fusion."

However, when presented with the facts that nothing was done on the subject for twenty-nine days after the meeting and that he had reviewed the Pons-Fleischmann proposal by then, Jones insisted that this level of activity still legitimately meets the definition of "vigorous pursuit." He did not deny that he may have had "impetus" from the Pons-Fleischmann proposal but argued that Pons and Fleischmann had not accused him of "impetus"—they had accused him of stealing ideas wholesale. Jones conceded that perhaps in drafting BYU's official account he should have noted that he had assigned a student to do electrolysis experiments (of the kind Paul Palmer had pursued two years earlier and Pons and Fleischmann were now proposing) only after reading the Utah proposal.

Ryszard Gajewski, who had unshakable faith in Jones, later offered a parable to justify why Jones may have believed he was working "vigorously" and "100 percent" on cold fusion. This was a story told to him by Victor Weisskopf, an eminent physicist at the Massachusetts Institute of Technology. "It is about the Chinese painter who was commissioned by the emperor to paint the most beautiful picture that was ever painted. He was given three years to do that, and the understanding was, if he did it he would be very amply rewarded, if he did not, he would be beheaded. So he was given a palace and servants, and just as three years were about to elapse, and there was no word from him, the emperor sent a delegation to the palace. There were just a couple of days left, and they wanted to see the painting. The painter showed it, and it was just blank canvas. The delegation went back and told the emperor. A couple of days later, the emperor and his court went to the palace, and there it was, the most beautiful painting that the world has ever seen. And the emperor asked how come two days ago there was nothing, what have you been doing for all that time, and the painter said, 'I've been thinking.' "

To this Gajewski added his own quasi-rhetorical question: would he be surprised to discover that Jones, consciously or subconsciously, intensified the pace of his cold fusion research because of what he saw in the Pons-Fleischmann proposal? He said he would be unable to answer definitively. "Maybe he did or maybe he didn't, but I would not be surprised if he did. I have no evidence to that effect. It's just human nature."

Whether he did or not was important merely because Pons and Fleischmann believed that Jones only "vigorously" began his research *after* reading *their* proposal, and that the fate of billions of dollars, among other things, hinged on whether he did or not. And what Pons and Fleischmann believed, rightly or wrongly, was what led them publicly and emphatically to disclose their invention on March 23, which is to say well before they had gathered sufficient data to support their claim.

The evidence, oddly enough, for Pons and Fleischmann's version of the events is persuasive. Indeed, Jones's logbook seems to bear out their version rather than his own. For that reason, the logbook became one of the more important witnesses in what followed.

The August 24 fusion group meeting, which figured so prominently in the BYU history of cold fusion, was attended by various of Jones's undergraduate and graduate students who were back on campus for the start of the 1988–89 academic year. Paul Palmer was there, as well as Gary Jensen, a professor of physics who worked with Bart Czirr on detectors. Jones recorded the event in his abbreviated fashion in his logbook:

> 8/24—Fusion group mtg. —electro-
> approach: wire—vary D_2
> : foils
> :gamma & neutron
> :$—this year /next
> : proposal

Jones later deciphered these notes this way: "We talked about electron-catalyzed fusion, as we were also calling it at that time. We were going to use various wires and foils, varying deuterium concentrations—this is cryptic as usual—look for both gammas and neutrons. And we needed to come up with some proposal. We needed to worry about money this year and next." Jones made no mention of electrolysis or an electro-chemical method of loading deuterium into metals, an omission worth a moment's consideration.

In one of the many coincidences in this affair, Steve Pons sent his proposal to the University of Utah's Office of Technology Transfer on the same day that Jones held his fusion group meeting. With the pro-posal, Pons was following a standard procedure and informing the uni-versity patent lawyers that he and Fleischmann had a potentially patentable idea and that, if a patent were to result from it, they now had a date of priority.[1] Or so they thought.

Pons then sent the proposal to Gajewski, who officially received it on August 31 and sent it on to five reviewers, two of whom were Jones and Rafelski.

Through September 20, Jones made no further mention in his log-book of piezonuclear fusion or electron-catalyzed fusion. On the twen-tieth he recorded that he had received the Pons-Fleischmann proposal and, apparently, discussed it with Gajewski. Now Jones was referring to his own piezonuclear fusion research as mineral-catalyzed fusion, which may or may not be relevant:

> 9-20-88
> —R. Gaj—U of U proposal:
> —Discuss mineral-catalyzed fusion!
> —work underway at BYU—since '86
> —part of cold fusion research—reported in
> '85–'86 annual report → suggest joint effort
> —Background radiation—what from?
> how much?

Gajewski mailed a cover letter with the proposal. According to the BYU account, that letter said "nothing about declining to review the proposal if the reviewer was doing related work. Indeed, most of the proposals which Dr. Jones is asked by the DOE to review relate closely to his active research on cold nuclear fusion, including muon-catalyzed fusion." Gajewski's responsibility as a funding officer was to have the proposal reviewed by the best-qualified referees, and undoubtedly he felt Jones fit that description. "I think if I did not send it to Steve," Gajewski said later, "I would have been criticized for not using the advice of a person who is perhaps the only person in the United States who is thinking and working along similar lines."

That Jones was preparing to launch a similar line of research may not have mattered, except for a second clause in Gajewski's cover letter, which stated that the reviewer agreed to "use the information contained in the proposal for evaluation purposes only."

This is where the issue turned. Jones certainly had doubts about following up on the review. "Possible conflict of interest," he had told Palmer. He also was immediately aware, or Gajewski made him aware, of Pons and Fleischmann as competitors, or potential collaborators, hence the note in his logbook "suggest joint effort." He received the Utah proposal and then remarked about the history of his own work. Perhaps Jones himself was simply stunned by the remarkable coincidence. Jones later said he reviewed the proposal because he trusted Gajewski's judgment, and until he did so he couldn't know how closely it bore on his work, although Gajewski could and did.

Gajewski, more than anyone, knew that Jones was in the market for a new line of research and that the JASON review would not save his program. It seems an unlikely coincidence that both Jones and Rafelski received the proposal from Gajewski in mid-September, just before a muon-catalyzed fusion meeting that Rafelski was hosting in Tucson. All three scientists would be at the Tucson meeting. Thus, they could, and would, discuss what research Jones and Rafelski would pursue next.

On September 21, the day after Jones noted his criticisms of the Utah proposal in his logbook, he flew to Tucson for Rafelski's meeting. That day he composed a list of evidence for the existence of piezonuclear fusion (which he again called mineral-catalyzed fusion).[2] At the end of the list, Jones wrote, "school starts—new neutron counter . . . nearly ready."

On September 22, in Tucson, Gajewski publicly revealed JASON's pessimistic critique of muon-catalyzed fusion research.[3] Although Gajewski said that he remained optimistic and that they were not dead yet, Jones wrote succinctly in his lab book, "critique 'Impossible.' "

At some point during the three days in Tucson, Gajewski, Jones, and Rafelski discussed the future of the BYU research program. Gajewski recalled that both Rafelski and Jones suggested they move back, as a team, into the piezonuclear fusion work, and Gajewski agreed. If the science was valid, it would be an interesting new approach to fusion. It would not, he said, "constitute a continuation of this muon-catalyzed fusion grant."

What was curious about all this was that, when Gajewski sent Jones the Pons-Fleischmann proposal, he was very much aware of Jones's interest in piezonuclear fusion.[4] In fact, he later said that when he first received the proposal he was "startled" at how closely the research resembled Jones's. He insisted, however, that he could not recall any discussion in Tucson pertaining to Pons, Fleischmann, or their proposal. "But I do remember," he said, "that Steve and Jan felt that there is a need to go ahead with this piezonuclear fusion experiment." The idea that both Rafelski and Jones received the proposal and then refrained from discussing it with Gajewski is difficult to swallow.

Jones at first contended that he had not read the proposal immediately, but that argument was not consistent with his logbook, which is to say his September 20 notes on the proposal. Rafelski said that because of the extra administrative burden of preparing a workshop, if *he* had received the proposal, he would not even have opened it until afterward. "I definitely did not have any intimate knowledge of the Pons and Fleischmann proposal at the time of this workshop," he said.

In any case, while still in Tucson, Jones made a list of things to do. It included a note to get copies of "piezonuclear fusion" to Gajewski, apparently referring to his 1986 paper with Clinton Van Siclen.

Another note read, "call Palmer, n/gamma detection, furnace/electro-chemical—direct students."

Two days later, according to the logbook, Jones and Gajewski shared a ride to the Tucson airport. Once again Gajewski said that it was time to wind down the muon-catalyzed fusion program; his support would last only until November 1989, unless Jones had a breakthrough or went off in a new direction. So Jones once again began to work on cold nuclear fusion, although still by proxy through his students.

The assignment went to an undergraduate chemistry major, Eugene Sheely. Sheely had worked for Jones since 1986, mostly on muon-catalyzed fusion, and he'd had his name on a few published journal articles. During the first half of September, Sheely had done related work at Los Alamos. On the seventeenth, he got married and took off to Las Vegas for a honeymoon.

When he returned to BYU on the twenty-seventh, one week, at least, after Jones had read and noted his criticisms of the Pons-Fleischmann proposal, Sheely received a note from Jones requesting that he start doing electrolysis and looking for signs of fusion. Stuart Taylor, a first-year physics graduate student, was assigned simultaneously to try some of Paul Palmer's electric spark approaches.

Jones also arranged with Bart Czirr to use his detector to observe Sheely's electrolysis cell.[5] Czirr then began setting up a gamma ray detector, because his latest neutron detector was still out of action. They would have preferred to look for neutrons from deuterium-deuterium fusion because the signal would be more definitive, but they had to use the equipment at hand.

Jones also discussed cold fusion with Daniel Decker, chairman of the BYU physics department. Jones showed him Palmer's two-year-old gamma ray spectrum, with its little bump that Jones thought might be evidence of fusion.

"Yeah, well," Decker said, "there might be something there. It is an awful small bump."

And Jones, as Decker recalled, said, "This is real. Can you help me figure out what is really going on?"

Decker then outlined on his blackboard an elementary model of the system. He modeled the electrode in Palmer's cold fusion cell as a lattice of charged metal ions, stripped of their electrons, then all these electrons flowing freely among the ions. It was as though this sea of electrons was washing around islands of metal ions. The deuterium atoms would enter the metal lattice, give up their electrons to this sea, and then, maybe, fuse.

Decker wrote down the equations for the forces between the deuterium nuclei in this sea of electrons. The nuclei would be pulled toward each other by the strong force, the same attractive force that binds the protons and neutrons within each nucleus. Simultaneously, they would be repelled from each other by electromagnetism, because the nuclei have an overall positive charge from their resident protons, and these like charges repel. The strong force, as the name implies, is much stronger than electromagnetism, but it fades quickly with distance.

The pivotal question was whether the deuterium nuclei could ever get close enough so that the repulsion of the electromagnetic force would be overpowered by the attraction of the strong force. If that could happen, the two nuclei would become one.

According to Newtonian physics, fusion would never occur under these circumstances. But quantum mechanics provides an out, which is to say it assigns every action a probability and, unless the action in question violates some fundamental law, such as conservation of energy, it would never have a probability of exactly zero. This uncertainty is the fundamental nature of quantum mechanics. In the case of fusion, there is always the possibility, however infinitesimal, that any two deuterium nuclei would "tunnel" under the barrier of their repulsive forces and fuse. This is what Jones had calculated in his paper with Van Siclen in 1986. Tunneling of deuterons is the quantum mechanical equivalent of a snowball's chance in hell.

What Jones hadn't calculated, however, was the crucial unknown. What about that sea of free electrons in the metal? Did it somehow manage to cancel the electromagnetic repulsion so that the deuterons could get close enough for the strong force to fuse them?

Jones had a graduate student copy the equations off Decker's blackboard, write them up as a computer program, then plug in the numbers. The result, Decker recalled, was that the probability of the deuterons fusing was, for all intents and purposes, zero.

But that was just theory. Steve Jones was an experimentalist.

On September 30, Jones sent Gajewski his review of the Pons-Fleischmann proposal. He suggested that it not be approved,[6] at least until Pons and Fleischmann could address a number of weak areas, which he enumerated. The last of these, Jones said later, was that Pons and Fleischmann had cited no references. He wrote, "One wonders if a thorough literature [search] has been done. In particular, publications by C. Van Siclen and S. E. Jones (J. Phys. G. 12 [1986] 213–221) and by B. A. Mamyrin and I. N. Tolstikhin (*Developments in Geochemistry 3: Helium Isotopes in Nature,* New York: Elsevier, 1984) could be relevant."

Jones said he suspected that mentioning his own obscure publication, coauthored with Van Siclen, would be a powerful hint that he had authored this anonymous review. And Fleischmann later remarked that both he and Pons knew from this first go-around that this review had come from Jones. He said of the five reviews, two were positive, two definitively negative, and one equivocal. The last was Jones's.

The piracy scenario that Pons and Fleischmann later constructed was based on their eventual, if not immediate, identification of the Jones review. At least, according to Chase Peterson, this is what they told him when the growing confrontation came to his attention. "You see, this proposal of Stan and Martin's is rejected. [The reviewers] say it's a worthless thing to promote because we can't see how it will work. So Stan and Martin think, 'Do we answer it? Do we tell them more?' As they report the story to me they said, 'Well, we guess we have to.' So they give some answers to the questions posed. And the second letter of opinion came back from Jones, still not understanding why it might work, and they finally gave him more data and more rationale, and then the response says, 'Oh I see, maybe it would be worth doing,' and that's what got Stan and Martin very concerned. They felt that until they had explained it in more detail Steve Jones was not aware of the mechanisms by which this could be done. Even though he tried it presumably, and he has these notebooks notarized and so on. But it had failed and he had pretty much abandoned it; then, only after these letters were exchanged, he tried again, and he started to get positive results."

Since Joey Pons had left in August to get married, the fusion project at the University of Utah had been in hiatus. Apparently Pons was waiting for a response on the proposal and for the money that would come if Gajewski agreed to support him. It was in late October, about the time Pons received the first round of reviews, that he asked Marvin Hawkins to take over the experiments.

Hawkins was old for a graduate student. He was twenty-seven when cold fusion entered his life. He was also the only member of what he would call "the local religious persuasion" in the Pons laboratory. He had taken a two-year mission in Australia, delaying his entry into graduate school. Hawkins had done his thesis research on microelectrodes and had proven an ability to acquire the requisite instruments, build the apparatus, configure experiments, and work with electronics. He was also an accomplished glass blower. To Pons, who was acutely concerned with secrecy, this meant that he could keep his fusion experiments running without recruiting help from the chemistry department support staff, who might talk. Pons even used Hawkins to purchase equipment so he wouldn't have to go through the usual university channels.

And Hawkins wouldn't talk. He was, at least until March 23 or so, devoted to Pons. He had grown up on a farm on the plains of western Colorado—"chasing cows and raising hay and enjoying the good life"— and found Salt Lake City too "big" for his tastes and the general population entirely too self-interested. Not only was Pons generous with his time but he was somebody to whom Hawkins could turn when he needed help. As for Fleischmann, Hawkins referred to him as a "god." Fleischmann's grandfatherly disposition, he said, had won his heart. Fleischmann later told Hugo Rossi that they gave the cold fusion assignment to Hawkins because "we knew we could trust him to do what we told him to do, [and] do it correctly."

Before Hawkins took up cold fusion, he had been feeling claustrophobic at the university, which was a symptom that he had been a graduate student long enough. After four years at Utah, he knew what he wanted. He observed that with a doctorate in chemistry, he could begin work at $40,000 to $50,000 a year. He was earning $10,700 as a graduate student in Pons's windowless basement laboratory. On that salary he had to support a wife and three children, which was not possible even in Utah. He was surviving with guaranteed student loans. "I was tired of being low on the totem poles," he later recalled. "I was tired of running. I wanted to finish and go on." Another reason Pons recruited Hawkins out of his dozen or so postdoctoral and graduate students was that Hawkins had just completed his thesis research, so all that remained was the writing. When Pons requested that Hawkins fabricate cells and take data on the fusion project, Hawkins figured the work might postpone his entry into the real world by as long as six months. He also considered the fact that the research could lead to a share in a Nobel Prize: "I'd have been an idiot to have said no." Once Hawkins took up cold fusion, he began to think that maybe it was his destiny.

When Hawkins took over the experiments in late October, his mentor's concern with nuclear radiation was still limited. Said Hawkins, "If we're getting zapped with heavy doses of radiation, we want to know about it." Once the reviews on his proposal came back, or at least about that time, Pons investigated obtaining some radiation detection equipment for his laboratory.

Pons called Robert Hoffman, a health physicist on campus, and asked what kinds of instruments he would need to measure neutrons, protons, or gamma rays.[7] The question suggests that Pons had not given much, if any, thought to the actual physics of nuclear fusion before that time. It also suggests that Pons was responding to his reviewers, who would surely have suggested that the standard evidence for the existence of a nuclear reaction is nuclear radiation—in this case, neutrons or gamma rays.

Hawkins, meanwhile, had spent November designing and running cells by trial and error. Pons had ordered and received four palladium rods to use as electrodes, and Hawkins was trying to create the perfect fusion cell for them. It would take him a week to put a practice cell together. Then, as Hawkins put it, he'd "try it, crash it, retrofit, change and redrill and rework, take maybe another four or five days, and try something different."

Hawkins said that until he began his trial and error program, they had been running only one fusion cell in the lab. This apparently was the one that Fleischmann would later refer to as their "monster cell." The monster's electrode was a sheet of palladium, eight centimeters square by two millimeters thick, which was the highest gauge commercially available. It was bent in a U-shape and placed in a large glass vessel with a platinum electrode in the center.

For a variety of reasons, Pons had not been happy with the monster, although they apparently kept it charged. When Hawkins joined up, Pons explained to him that he wanted a cell that would allow him to monitor and control carefully the temperature emitted by the cell, while distributing a uniform current around the electrode. The monster apparently failed to fulfill both of these requirements.

Hawkins said his first cells were beset by a variety of problems, the most stubborn of which was a tendency of the palladium electrode to break or short out as it absorbed deuterium and expanded. Hawkins beat this by building a cage of four glass rods around the palladium rod, then winding the platinum anode around the glass cage. This entire assembly fit inside a double-walled glass vessel, known as a dewar flask, which was sealed off at the top by a plastic stopper. (Hawkins's account implies that the infamous fusion cell, which later would be imbued with nearly magical powers, had been designed by Hawkins himself, a graduate student who had worked by trial and error with little guidance from either Pons or Fleischmann.)

Forty-five miles down the road in Provo, Steve Jones was neither waiting to throw himself into cold fusion, as Pons and Fleischmann would later surmise, nor deviously plucking the two chemists for details. He was simply letting Eugene Sheely, his student, work on it while Bart Czirr and Gary Jensen helped with the detectors.

Sheely had read up on electrolysis and asked James Thorne, his adviser in the chemistry department, about it. He said he found it not particularly difficult to learn. He spent a few days procuring the necessary equipment and the heavy water. He used sodium sulfate for his electrolyte, because he had it in the lab. He assumed any salt would do, even table salt, as long as it didn't corrode the electrodes. And Thorne provided a power supply and suggested he use stainless steel electrodes.

When those turned out to rust quickly in the electrolyte, Sheely switched to nickel and eventually palladium because, he said, Jones suggested several times that he try palladium.

Why palladium? Jones insisted it had nothing to do with what he might have read in Pons and Fleischmann's proposal. According to Jones, Thorne suggested palladium as the natural choice because it absorbed so much hydrogen. At any rate, Jones included it in the list he had written down back on April 7, 1986, as a metal that "dissolves much hydrogen."

That was one way of looking at it. Another was that after reading the Utah proposal Jones did indeed discuss this issue with a chemist, Thorne, and then assigned a chemist to the electrolysis instead of one of his physics graduate students. This train of events was consistent with the pirating scenario that Pons and Fleischmann later constructed. Jones, the physicist, seemed to be pursuing cold fusion from an electrochemist's perspective. Impetus, to use Jones's word, appeared to be present.[8]

Either way, Sheely eventually got his cells running but spent most of his time waiting for Czirr and Jensen to get the gamma ray detector running as well. The device seemed to be in a constant state of disrepair.

Jones came down to see the work a couple of times, and he had a group meeting once a week, during which he always asked Sheely for a progress report, but he seemed to be preoccupied with the problem of getting money for muon-catalyzed fusion. Even after the Tucson meeting, according to the record in Jones's lab books, both Gajewski and Jones were still determined to find money to save the program.

On November 8 Jones spoke at length to Gajewski about muon-catalyzed fusion. Jones wrote in his notes that Gajewski had looked into DOE's conventional fusion programs as a possible home for it, but he had "little hope." And DOE's Atomic and Molecular Physics Division, which would also be a possibility, was "clogged" with projects and had "little turn-over."

Finally Jones and Gajewski discussed what is known disparagingly in scientific circles as earmarking, appealing directly to Congress for support, thus bypassing entirely the DOE bureaucracy, peer review, and all the attendant frustrations. Jones sketched out a possible plan of attack: Define the best congressional committee, approach the committee, choose one of his local representatives to go through, and so on. With luck and perseverance, he could end up with a bill in Congress that said that so many dollars had to be spent by the Basic Energy Sciences Division of DOE on muon-catalyzed fusion. Still, earmarking was considered a last-ditch contingency.

From November 9 onward, Jones began to attend personally to piezonuclear fusion. On November 16 he wrote in his logbook, "pons response," among a list of things to do, and he checked it off. This was his first mention of Pons in his book since September 29. Apparently, Jones had now sent in his review of Pons's revised proposal. Jones said later that this was the review in which he recommended that the Pons-Fleischmann proposal be approved.

On November 28 Jones spoke to Rafelski, who apparently told him that Pons "needs nuclear physics help." Jones noted that "Jan wrote of their ignorance, rejected proposal."

The next entry in the logbook is December 1, and it finally concerns Sheely's electrolysis work.

Sheely had now given the results to Jones in an organized form. On top of the page, Jones wrote, "Pd + H_2O + D_2O runs show small excess of neutron rate vs background." Sheely later translated this as "exciting but definitely real close to background. Nothing anyone thought was publishable, we needed more data, but it was enough that it was more than worth continuing to test it."

Then come several pages of calculations working out the signal and the background from Sheely's data. All the runs seem to have been recent and Czirr's neutron detector, it seems, was finally running. There are four pages of these calculations, the last of which is dated December 5.

On December 9 Jones was in Washington, meeting with Gajewski and Rafelski. Once again, Gajewski ran down possible sources of money for muon-catalyzed fusion, and Jones actually drew in his logbook a schematic diagram of the DOE hierarchy. Evidently the directors of the conventional fusion programs had declined to offer support for Jones's work. But the division head at Atomic and Molecular Physics would accept a short preproposal. Jones wrote, "Congressional route a possibility." Also possible were the National Science Foundation and EPRI, the privately owned Electric Power Research Institute.

Then the discussion turned to cold fusion. Jones described Sheely's results. Then comes this note, at the top of page 59:

12-9-88   stimulating nuclear fusion by means of flow of
                    hydrogen isotopes in metal lattice

/

Jan: 1—Patent: Jones, Rafelski, Palmer

Rafelski apparently suggested that they consider patenting the discovery of "stimulating nuclear fusion by means of flow of hydrogen isotopes in metal lattice" and that the names on the patent should be Jones, Rafelski, and Palmer.[9] This is a remarkable suggestion, especially when made by two reviewers of a proposal on the identical subject that, if nothing else, had provided an "impetus" to pursue their work. The names of Pons and Fleischmann are conspicuously absent here. As none of the principals professes to remember details of the conversation, this is the only record that exists.

Gajewski then told Jones that he could also submit a proposal on cold nuclear fusion, independent of the Pons-Fleischmann proposal.

On the flight back to Salt Lake City, Jones composed a draft of what was either his piezonuclear fusion proposal or a paper claiming its discovery. Jones does not remember which, he only remembers writing it.

The title is "First Demonstration of Cold Piezonuclear Fusion." The six pages begin with Jones's work in muon-catalyzed fusion and describe how that led them to believe that there must be other ways to induce hydrogen atoms to fuse. After discussing the geophysical evidence for such fusion, the draft sketches a quasi-theory that would allow hydrogen nuclei to get close enough to fuse in a metal lattice provided two hydrogen ions could inhabit the same site. This, Jones wrote, would be enhanced "when the ions are rapidly migrating in the host metal, jumping between interstices." Thus, with what could be called either vigorous hand waving or a miracle, "cold piezonuclear fusion in metals (one could call the process metal-catalyzed fusion) appears feasible at a very slow rate."

These considerations, Jones continued, "have motivated a series of experiments at Brigham Young University to search for metal catalyzed fusion." He then described Sheely's recent electrolysis experiments and wrote that they had seen both gamma rays and neutrons with a "specially-designed neutron counter [that] was built for this experiment."

He ended with his characteristic optimism:

In conclusion we have demonstrated for the first time that nuclear fusion occurs when hydrogen and deuterium are electrolytically loaded into a metallic foil. This remarkable process obviates the need for elaborate machinery to generate and contain either plasmas or muons to induce fusion. We are now exploring means to enhance the fusion yield of this new process.

At this time Jones was still considering two strategies. At some point after receiving the first version of the Pons-Fleischmann proposal, he suggested that Gajewski inform Pons outright of the existence of Jones and his research. Jones said he may have made this suggestion immediately upon receiving the proposal, as early as September 20, which his logbook bears out. Yet Gajewski didn't act on it until December 16, after getting the second round of reviews back, after Jones had time to get the results from Sheely's electrolysis experiment and discover that there might be something to this phenomena, and after Jones, Rafelski, and Gajewski met in Washington to discuss the future of cold fusion.

Jones later admitted that he considered keeping his research secret and going public first. This gambit appears several times in his lab books between December and March. "There's this little closet under our stairs, carpeted," Jones said, "and I'd go in there and I would think, and I would pray, and I would think, What do I do? How do I handle it? I've read their proposal. I've been working on this, and I don't deny there could have been some impetus. And the solution I came up with was tell them and let them use our detector."

Gajewski apparently favored a similar strategy. The way he saw it, Jones had the physics expertise and the radiation detection equipment. Pons and Fleischmann had the chemistry expertise. They were attacking the same problem from distinct but mutually supportive directions. "Pons was working on heat," Gajewski said, "and Steve was working on neutrons. And I would have been delighted at the time to fund both, one to continue working on neutrons, the other to continue working on heat, and possibly collaborating and checking on each other."

It made perfect sense, except that at this time neither had any data worth mentioning. Still, according to Jones, Gajewski even took into account "that a cooperative effort from the nearby schools could provide an efficient use of taxpayers' money if things worked out and their proposal was approved." Bart Czirr wanted to sell Pons and Fleischmann one of his detectors, and he understood, as did Paul Palmer, that Gajewski was telling Pons and Fleischmann that they wouldn't get funded without a decent neutron detector. "Gajewski told them to get down here and learn how to measure neutrons," said Palmer. "They couldn't get funded in nuclear physics if they didn't know how to measure neutrons. Gajewski, in my opinion, was doing the best job a civil servant could do. He said, 'Why don't you stupid guys cooperate?' [To us, he said,] 'You need electrochemical expertise. You're doing electrochemistry, and you're doing it ignorantly and stupidly. Why don't you get some electrochemists?' And he told them, 'You guys don't know anything about nuclear physics. Why don't you get some nuclear physics help? Some detectors. Why don't you work together?' "

Finally, on December 16, comes this note in Jones's log:

> R. Gajewski—spoke to Pons, who had my rebuttal & Mamyrin paper
> —benefit to science aided by interaction
> but no pressure to work w/ Jones
> —get together & measure each other
> —spoke of '86 report to him (Jones—mcpf)[10]

And later that day, Pons and Jones apparently spoke for the first time:

> Pons: several weeks to load p-d to start expt.
> interested in our neutron detector! seeks help

According to Jones, he then sent Pons the requested information about Czirr's neutron spectrometer, and Czirr sent along one of the proposals he had drafted for the Department of Energy. Czirr recalled that he had carefully chosen this material so that Pons and Fleischmann could see what he had to offer but not use the information to build a detector themselves.

Although from Jones's notes Pons seems to have been amenable to a working relationship, that was not the case, even this early in the game. If Pons already believed he was sitting on the discovery of a lifetime, how was he expected to take these two phone conversations? Pons would later say these calls were the first time he ever heard of Steve Jones, and the BYU physicist was immediately suggesting they work together.

Gajewski may have been the first funding agent to whom Pons showed his proposal who could have been considered a stranger of any sort. Up until Gajewski, Pons had been reluctant to tell his *friends* about cold fusion. Even when he told Richard Bernstein of UCLA about fusion in September 1987, he swore Bernstein to secrecy.

Now Gajewski called him and suggested that he collaborate with one of his reviewers. Gajewski also told him of the report that Jones had filed in 1986 on piezonuclear fusion, which is to say cold fusion.

Gajewski "spoke of '86 report to him" is how Jones recorded it in his notebook. So Gajewski, whether he realized it or not, was now telling

Pons that Steve Jones, his reviewer, had been doing the very same work since the spring of 1986. (Quite a coincidence! So Jones, the reviewer, was first.)

Stan Pons could not have been pleased. He was alone; Fleischmann would be in England for quite a while. And he was being robbed. Although Gajewski suggested that Pons collaborate with Jones, it must have appeared to Pons as though Jones and Gajewski had decided that they were moving in on his territory. It seemed like a fait accompli, which in a sense it was.

It was around this time that Pons met Eugene Loh, the chairman of the Utah physics department, at a campus Christmas party. Pons asked him where to obtain a neutron detector and actually described his experiment. Loh later said the concept seemed so ludicrous that he assumed Pons was joking.[11] As for the neutron detector, Loh directed him to the Los Alamos National Laboratory. "Pick up a phone," he said. "The people at Los Alamos will be glad to lend you a detector."

Instead, Pons seems to have called back Robert Hoffman, the health physicist, who lent him a small portable detector. This was a dosimeter, which would effectively do nothing more than alert Pons to the possibility that he was getting zapped. "If that thing ever went off," Marvin Hawkins said, "we knew it was time to shuffle on out." Of course, if his electrochemical cells were inducing any appreciable level of fusion, as Pons believed they were, the dosimeter should have been more than sensitive enough to pick it up. It registered nothing.

By the Christmas holidays, Hawkins had finally constructed cells that he considered "reasonable." Both he and Pons worked through the vacation. As Hawkins recalled it, they were in the lab "all day every day through this period of time."

Pons and Hawkins were now ignoring the issue of nuclear radiation and concentrating on the first of endless measurements of the heat generation from the cell, a procedure known as calorimetry. The purpose of these measurements was to compare the energy going into the cell with the energy leaving it. This required, among other things, constant monitoring of the current and voltage feeding the electrodes and the temperature of the electrolyte in the fusion cell. If carried out correctly, the procedure would show whether the energy emitted by the cell was greater than the energy it took to run it. If this were the case, as Pons, Fleischmann, and Hawkins would come to believe, then the excess energy could be attributed to some energy generation in the cell, perhaps some unknown species of radiation-free nuclear fusion.

Richard Steiner, associate chairman of the chemistry department, said

Pons had talked to him about calorimetry and how difficult it is to do it right. "He talked to me," Steiner said, "about having read this book and that book and looked at this method versus that method, and worrying about how to do it." Fleischmann, however, later told Dave Williams, an electrochemist at the Harwell laboratory, that he had a foolproof method for doing calorimetry. As it would turn out, there is nothing simple or foolproof about calorimetry.

In the organized competition to contribute to man's scientific knowledge, the race is to the swift, to him who gets there first with his contribution in hand.

ROBERT MERTON,
The Sociology of Science

*B*y early January 1989, Steve Jones was finally dedicating the better part of his time to piezonuclear fusion. Now his own words, such as "vigorously pursuing" or "running as fast" as he can, could almost be deemed appropriate.

Bart Czirr also had his neutron detector running, albeit not well. Gary Jensen, Czirr's partner, said they switched from the gamma ray detector to the neutron detector only because the background noise picked up by the gamma ray detector made it impossible to observe anything.

Detecting neutrons, however, was a perverse endeavor in its own right: on the one hand, neutrons have no electromagnetic charge, which makes them nearly invisible to most detector schemes. On the other, the world is full of naturally occurring neutrons from cosmic rays, known as background radiation. In order to prove the existence of cold fusion, Jones's experiment had to differentiate between the natural abundance of neutrons in the laboratory without any fusion cells running—the background—and the abundance of neutrons in the lab when the cells were running—the signal. If the signal was greater than the background, then maybe they had something.

Czirr's neutron detector was a clever modification of a standard design for such devices. It used a material called a lithium glass scintillator, which, in this case, consisted of plates four inches square and a few millimeters thick. If a neutron passed through the glass, it stood a good chance of being captured by an atom of lithium. The lithium would then emit an alpha particle—a helium nucleus—which would have a positive charge, and this would create a pulse of light as it passed through the glass. This light pulse would be detected and amplified and would eventually register as a neutron.

Since 1986 Czirr had been modifying the device so that it could also measure the energy of these neutrons, which could be valuable information. Deuterium-deuterium fusion, for example, always results in a neutron of exactly 2.45 million electron volts (MeV). If Czirr's detector could measure the energy of a neutron, this would simplify the chore of proving whether the neutron came from fusion or some background source. Czirr said the resolution of his detector still fell considerably beneath his expectations: a 2.45 MeV neutron, for example, might show up anywhere between 1.00 MeV and 5.00 MeV.

Czirr would have preferred to spend his time working on this resolution problem, but instead he was continually struggling to keep the neutron detector and its attending electronics operating. "We had old borrowed equipment," Czirr said. "It's a bunch of junk really. Half of it, anyway. So it was hard to keep it operating and not running out. So we had our eyes focused on the detection of neutrons and the detector itself. Is it working or is it not?"

Jones, his various students and colleagues were still working at what constitutes a leisurely pace in the research community. When Jones eventually got around to publishing, his data consisted of fourteen experimental "runs," all between December 31 and the first week of March, and lasting from eight to twenty hours each. This works out to maybe two runs per week. So they were running cells every second or third day and regularly changing the electrodes and the mixture of the electrolyte, hoping for some combination that would produce copious neutrons. Now and then, they would leave Czirr's detector on at night without any cells running to measure the neutron background.

Jones assembled the data almost daily. "He was watching it all the time," said Czirr, "seeing if there was any hint of a signal sticking its nose up above the background."

They had intimations on occasion, but nothing that a physicist would consider compelling. On January 5, for instance, Jones wrote that they had observed a signal using a titanium electrode and sodium sulfate for the electrolyte:

$$Ti + d + n + Na_2So_4 \text{---very high n° rate}$$
Fusion neutrons are highly suspected!

A week later, Jim Thorne, the chemist, suggested that they try adding metal ions to the electrolyte. This procedure is known as poisoning, which would help the palladium absorb hydrogen and deuterium and might increase the fusion effect, if there was one.

During these first weeks of January, Jones was also writing up his cold fusion proposal, and Jan Rafelski seems to have been spurring him to submit it quickly. On January 10, for instance, Jones wrote in his lab book that he spoke with Rafelski, who suggested an aggressive strategy:

Jan:
Proposal—submit urgently—on basis of own findings
—valid project

That the project was valid seems reasonable, but it is hard to understand what Rafelski and Jones considered the pertinent findings to have been on January 10. They still had no neutron signal worth taking seriously. They did have Paul Palmer's geophysical data, and by several accounts it was those data that convinced Jones that he had discovered cold fusion and thus should proceed with the proposal.

Until January, Palmer's evidence had depended on excessive amounts of helium 3 that had been found here and there but that could be explained away without invoking cold nuclear fusion. Palmer always thought that if they were to discover excessive amounts of tritium rising from the earth, that would be much more definitive. Tritium is a radioactive isotope of hydrogen, and, unlike helium 3, it has a very short life span. The half-life of tritium, which is to say the time it takes half of any given sample to decay radioactively, is only a dozen years. If the earth happened to be oozing tritium in excessive quantities, professional geologists could not explain it away as primordial residue, as they could with helium 3.

As Palmer recalled it, one of the graduate students, Stuart Taylor, located at the library a government publication by H. G. Östlund and A. S. Mason of the University of Miami, Florida, called *Atmospheric Tritium 1968–1984*. Among the voluminous data recorded by Östlund and Mason were the records of a monitoring station on the slopes of Hawaii's Mauna Loa volcano, which had recorded atmospheric tritium concentrations from mid-1971 through the end of 1977.

Palmer discovered in this data a near doubling of the tritium level during February and March of 1972. This happened to coincide with an eruption of the Mauna Ulu volcano, just twenty-five miles upwind. Later in the year Ulu erupted again, and again the tritium level increased, although not as dramatically.

To Jones and Palmer, the Mauna Ulu tritium seemed like a godsend, but the data were not quite so cut and dried. The picture was complicated, for instance, by atmospheric H-bomb tests, which had been blowing copious amounts of tritium into the atmosphere for forty years. And nuclear submarines had been flushing their fair share into the oceans. Nuclear reactor facilities and nuclear weapons facilities—like that at Savannah River, South Carolina, which made tritium for the H-bombs—had also been known occasionally to leak tritium into the atmosphere or groundwater. When this happens, neither our government nor those of the other nuclear nations are likely to advertise it, so it was doubtful that these technological sources could be ruled out as the source of the Mauna Ulu tritium. Man-made tritium was migrating through various ecosystems in startling quantities.

The Soviets had exploded a few hydrogen bombs underground just five months before Ulu's first eruption, and the United States did the same a few weeks later. Nonetheless, Östlund and Mason seem to have believed that these explosions could not account for the Mauna Ulu tritium, which gave Jones and Palmer more faith than ever in cold fusion. Jones later said that they checked with the U.S. Navy, which reported, of course, that no nuclear submarines were leaking tritium during that time. And Jones believed them, so that was that. "I'll tell you," said Stuart Taylor, "Steve was excited when he saw that tritium data."

As Jones would later put it in the first draft of his cold fusion paper, "We conclude that this volcanic eruption freed tritium produced by geological nuclear reactions."[1]

On the weekend of January 28, Dave Williams stopped off in Utah on his way back from New Zealand. Stan Pons told him that they were working on something "very hush-hush." Williams remembered that back in November Martin Fleischmann had told him "that he had this really interesting thing which was of potentially mind-blowing importance."

While he was at Pons's lab, Williams noticed that Pons and Marvin

Hawkins were busy unpacking a scintillation counter, a device for measuring tritium levels. Hawkins had already taken samples of the electrolyte from the fusion cells and had them checked for tritium at a radiobiology lab and at another chemistry lab. The results had seemed promising, so Pons had ordered his own scintillation counter. Pons and Fleischmann later reported that they had observed one hundred disintegrations of tritium per minute per milliliter of heavy water, which means that one hundred atoms of tritium were decaying each minute from a one-milliliter sample. This corresponds to a total of 1 billion tritium atoms in each milliliter, a misleading number because it sounds impressively large.

Heavy water, because of how it is produced, is likely to contain two or three or five times as much tritium as this naturally.[2] Pons and Fleischmann seemed unaware of this fact and considered their tritium data more evidence for fusion. After all, tritium can be created only in a fusion reaction. They were worried, however, because this number was still a billion times less than what they were expecting, considering the amount of heat that the fusion in their cells appeared to be generating.

Until the tritium counter arrived, Pons had managed to keep his fusion project successfully hidden from his other researchers. They knew something was going on, but they didn't know what. Afterward, they started hearing the word *fusion*.

By January 27, Jones was convinced that he had sufficient evidence of cold fusion to commit himself, and he debated a range of strategies. He could send Ryszard Gajewski his proposal, leaving Pons and Fleischmann to their own fate. He could write up his paper and appropriate the discovery. Or he could take the more charitable route and include Pons and Fleischmann on a joint paper or a joint proposal or both.

If he chose the aggressive strategy, as he noted on January 27, it would have to be with a quick and dirty publication of his quick and dirty experiments. Jones realized that this strategy involved taking a stand on his meager neutron data while admitting that he had little understanding of what he was seeing or why. He seemed aware that the geological evidence was still a flimsy foundation on which to build a theory of fusion.[3] ("Claim open Q[uestions]s" he writes.)

In late January, Jones sent a draft of his own cold fusion proposal to Rafelski for his comments, and then to Gajewski. On January 30 Jones noted in his logbook that Rafelski suggested he mention in the proposal his "early and continued interest." The intent is cryptic. One likely meaning is that Rafelski was less sanguine than Jones about the brewing conflict of interest. Once Jones had received the Pons-Fleischmann proposal, it could be argued that he was, in effect, damned. Any action on his part after that involving cold nuclear fusion could spark Pons and Fleischmann to accuse him of scientific thievery. He would have to scramble to defend himself. Gajewski later insisted vehemently that there was no conflict; the two proposals were not on the same subject. "Pons and Fleischmann were doing heat," he said, "and Jones was doing neutrons." But this point was questionable as long as both proposals involved electrolyzing heavy water with palladium electrodes to induce nuclear fusion.

Gajewski kicked the first draft of the proposal back to Jones, directing him, among other things, to get an electrochemist to help with the chemistry. Gajewski apparently hoped that Jones would go back to Pons and Fleischmann, looking for a collaboration. Instead Jones enlisted the support of Douglas Bennion, a prominent electrochemist on campus.

As Jones recalled, Gajewski also began urging him to go public with the data, saying, "It will be better if you publish first and then it will go more easily." Both Czirr and Palmer also picked up this impression along the way, although it's likely they got it from Jones. Palmer even came to believe that Gajewski said he could approve their proposal without an external review if they would publish a paper.[4]

Gajewski would later contradict this. In his version of events, Jones asked him if he should publish, and he merely suggested that if Jones was ready to publish he should. "Obviously it would help," Gajewski said later, "but at the same time, that was not my motivation. My motivation was that if he had something, there is no point sitting on the data because a guy in Singapore may be working on the same thing. That's elementary in science."

This seems a bit disingenuous. Gajewski surely knew of a guy or two in Salt Lake City working on the same thing. Either way, Jones took this suggestion as a mandate to announce his results.

Jones had already been invited to speak on muon-catalyzed fusion at an American Physical Society meeting in early May—the meeting of the year for the physics community. It is usually held near enough to Washington so that the scientists can visit their funding agents, interested politicians and bureaucrats, and suitably impress them.

On February 2, one day before the deadline for submitting abstracts for the meeting, Jones sent his by overnight mail. He briefly described the BYU work on muon-catalyzed fusion, then added just enough on cold fusion to constitute an announcement of discovery: "We have also accumulated considerable evidence for a new form of cold nuclear fusion which occurs when hydrogen isotopes are loaded into materials, notably crystalline solids (without muons). Implications of these findings on geophysics and fusion research will be considered." Employing an abstract to claim a discovery has a long and illustrious history in science. For instance, Edmond Halley, for whom the comet is named, urged Isaac Newton to do so in order to secure an invention to himself "till such time as he would be at leisure to publish it."

Jones later claimed that this was not what he had in mind. But when he was told that it was this abstract that partially led Pons to believe he was not dealing in good faith, Jones responded with an ambiguity worthy of a candidate for public office. "You've seen the abstract," he replied. "It was nothing. Yeah, 'We've seen evidence for this. . . .' I think it was kind of funny."

About this time Jones accepted an invitation to lecture on muon-catalyzed fusion at Columbia University on March 30. Amiya Sen, a plasma physicist at Columbia who had invited Jones, recalled that when he called him the conversation had a curious twist: "[Jones] said, 'By the way, I'm working on a little something else, and I'd like to talk on that also.' I said, 'Oh fine.' Then he said, I think, 'There is a competitor hot on my tail.' "

On February 3, Lee Phillips got a call from Norm Brown, head of Technology Transfer at the University of Utah. Phillips was Brown's counterpart at BYU, and the two had a collegial relationship. Brown said he needed to talk to Phillips "about the possibility of one of your professors' pirating one of our professors' stuff. What do you know about it?" Brown had been talking to Pons about the chemist's suspicions and assumed he could look into the matter with a simple phone call. He was still unaware that nothing in cold fusion would be simple.

Phillips told Brown that he knew little about the matter, and then asked Carol Hardman, who was responsible for overseeing BYU's outside research contracts. As it happened, the day before, Jones had informed her that he had submitted a proposal to the Department of Energy but that his funding agent had asked him to broaden the scope and resubmit it. Jones also mentioned that some chemists at the University of Utah were doing similar work, which seemed innocent enough.

Hardman assumed that Jones was debating whether to include the Utah chemists in the proposal. She relayed to Phillips this misconception of the issue, which he relayed to Brown: Jones had been working on cold fusion for a long time, he said, and the BYU administration had been keeping an eye on it.

This seemed to satisfy Brown. He said he'd call back if they "needed to get together on it."

Hardman then called Jones and told him that Norm Brown from Utah had been asking about the provenance of his proposal. Jones, concerned that Pons was spreading unjustified accusations, called Gajewski, who said he'd look into it.

In the meantime, on February 8, Jones met with a BYU patent review committee headed by Lee Phillips, regarding a "device to Produce Controlled Nuclear Fusion." Jones and Palmer presented this reactor-to-be, according to the minutes of the meeting, as though there was no prior work in the field, other than that by a "Russian author." The minutes continue: "Possible applications would be plastic explosive detection and a small energy source, i.e., for spacecraft. It is anticipated that a neutron source is still 3 to 5 years away. Work is in the early stages. No device is fully developed."

The meeting concluded with Phillips, Jones, and company agreeing that Jones should inform DOE that he will retain patent rights on the invention—as required by DOE—at which point they could take up to two years to decide whether to actually proceed with the patent applications: "The technology has unlimited potential, but it will still take many years to produce a marketable product. The technology is so technical that patent costs will be high, possibly $15,000 to $25,000. More information needs to be obtained before a decision can be made." That Jones, or somebody in this meeting, thought the technology had unlimited potential may be taken as indicative of how premature Jones's data still were. In any case, the patent review committee then decided that the only remaining concern involved waiting to apply for the patents themselves: "That is, should someone file a patent application claiming the matter which you may have priority to, it would be more difficult and expensive to fight the application than it would have been to file ours first."[5]

On February 10, Gajewski finally reached Pons about the Norm Brown phone call of a week earlier. Pons then talked to Jones directly, which appears to have been the first time the two had spoken since December. Pons apologized for Brown's intervention. He said they had "an overeager lawyer" at the university who had overstepped his bounds.

Then either Pons told Jones or Gajewski told Pons—Jones couldn't remember which, and his notes were vague—that Gajewski was going to hold on to Pons's funding until the conflict was sorted out. "I think," said Jones, "that Pons didn't like that."

In any case, Jones made no mention in his logbook of any talk of collaboration.

On February 14, Fleischmann visited the Harwell laboratory, where he was a consultant. Harwell is the largest applied research laboratory in Britain, and is often compared with the Los Alamos Laboratory in the States. It is one of the principal laboratories of the British Atomic Energy Authority and has three working fission reactors. Fleischmann's visit was specifically to procure the kind of nuclear physics help that Jones was offering.

Fleischmann had a private conversation with the research director of the lab, Ron Bullough. Like Fleischmann, Bullough was a fellow of the Royal Society, but, unlike Fleischmann, he was a professional physicist. He recalled that Fleischmann was genuinely concerned about the security aspects of his cold fusion invention. He told Bullough about the meltdown and added that they had seen nuclear signatures—in particular, tritium and neutrons—from other working cells. Fleischmann said that they had to be very careful how they handled it. "He felt he might be wrong," remembered Bullough. "There was an undercurrent of 'This is incredible and I have to be careful.' "

Bullough had trouble believing what he was hearing, although he took it seriously enough. "Given the man's background and his talent," Bullough said, "to say he was seeing nuclear signatures out of an electrolytic cell was, to put it mildly, astounding."

Fleischmann gave Bullough a draft of the Department of Energy proposal and asked him to discuss it confidentially with his physics colleagues. He then asked if he could borrow a neutron detector, and Bullough agreed. The Harwell staff took a detector, known as a Bonner sphere, off their production lines and sold it to Fleischmann for a few thousand pounds. ("I thought we had lent it to him," Bullough said later. "I am slightly embarrassed about that.") Finally, Fleischmann suggested that he and Pons might need help in confirming their results, and Bullough readily agreed to that as well.

Although it was left unsaid, Bullough believed that Fleischmann had also confided in him to give his country enough advance notice to reap some of the benefits of cold fusion. Fleischmann was anxious that the Americans not monopolize the technology. He wanted Harwell, and England, to be prepared to move fast should he and Pons have to go public.

Over the next few days, Bullough studied Fleischmann's proposal and discussed it with the head of theoretical physics. "I have to say," Bullough recalled, "there was complete incredulity."

Fleischmann arrived in Utah with the Bonner sphere a week later.[6] The Bonner sphere was a crude instrument designed for health physics purposes only. It was covered with a layer of plastic and was about the size of a human head, thus simulating the surface area of a human in proximity to a radiation source. If Pons and Fleischmann's cold fusion cells had been producing trillions of neutrons each second, as they should have if they were generating watts of fusion energy, then the device should have easily been sensitive enough to detect it. It registered nothing.

On Thursday, February 23, Fleischmann and Pons finally took Steve Jones up on his two-month-old invitation to visit Brigham Young University and look at their neutron detectors. The two chemists got a tour of the BYU experiment. Whatever Pons and Fleischmann had in mind with this visit, it came off as an acute exercise in miscommunication.

Jones and his colleagues had still been unable to detect enough neutrons to prove that they had generated fusion in their cells. When they went public one month later, it was with a reported signal of two neutrons per hour, which represents roughly one trillionth of a watt of power. Pons and Fleischmann, of course, believed that they had generated enough nuclear fusion to melt a cube of palladium. They believed they were producing untold watts of power. If so, they should have been seeing trillions of neutrons per second, but, according to their Bonner sphere, they were not. By the end of the day, as Dan Decker, head of the BYU physics department, recalled, "it was obvious that we had no idea really what they had done, and they didn't quite realize the minusculeness of what we had found."

The visit left the BYU group with a smattering of impressions. Decker, Czirr, and Palmer all remarked on the fact that Pons had said remarkably little the whole day. Czirr added that he found Pons quite likable. "I have something of a prejudice for southerners," said Czirr, who was born in Oklahoma. "He kind of made me feel at home with his twang." Fleischmann, in contrast, revealed an impressive knowledge of neutron detection, although how deep his understanding went was unclear. "He said all the right words," recalled Palmer.

Fleischmann seemed to be hoping that their absence of neutrons could be explained by some loophole in the esoterica of nuclear physics. He thought maybe Jones and his colleagues could, wittingly or unwittingly, help point it out. Fleischmann knew that when two deuterium nuclei fuse the result, in conventional nuclear physics, is either tritium and a proton or helium 3 and a neutron. The probability of either was about fifty-fifty. Fleischmann repeatedly asked Jones and his colleagues whether this branching ratio, as it is known, could be altered dramatically by the presence of the palladium lattice, which might explain why they had detected some tritium, albeit not nearly enough, and very few or no neutrons. Jones assured him that the branching ratio appeared to be immutable. The fifty-fifty split that nuclear theory predicted was essentially what had been measured in every fusion experiment, including muon-catalyzed fusion.

Pons and Fleischmann were disappointed in the answer. Neither seemed willing to doubt their conclusion that they had induced fusion, at least not in front of the competition. Perhaps their particular brand of nuclear fusion worked in a manner unknown to physicists. Certainly, they had fusion; they simply lacked the conventional evidence—neutron radiation. "They weren't dead, for example," observed Czirr, who knew well enough about the lethal nature of exposure to excessive neutron radiation. "Although I don't think they appreciated that point at that time."

The two chemists had brought along one of their own cells, and the BYU physicists were chagrined at how much larger and more professional than theirs it was. After seeing it, Czirr took to referring to the BYU cells as "the little, dinky baby-food jars." Fleischmann remarked that this cell was just a prototype. It had never produced neutrons, he said, which seems as good a way as any to have forestalled suggestions that they test it then and there.

Jones then revealed his neutron spectrum and the infinitesimal bump, which he considered his fusion signal. Pons and Fleischmann did not say what they thought of this evidence. "Oh yes, we showed it to them," Jones said of his spectrum. "We showed them a bump in the neutron spectrum. It's a very tiny effect, and I don't think that sunk in with them."

In fact, Fleischmann did realize it, although he may have construed that the effect was infinitesimal because Jones didn't know how to construct a fusion cell properly. One week after the BYU outing, Fleischmann called Dave Williams at Harwell and told him that they had spoken to Jones, "who [had] showed them this absolutely grotty neutron spectrum," as Williams put it.

After the tour of the laboratory, it was off to lunch at the Skyroom, BYU's rooftop faculty club. Now each side's misinterpretation of the two experiments began to play like farce. Fleischmann, for instance, warned the BYU physicists about the possibility of meltdown. Said Decker later the remark "went right over us."

The BYU physicists then asked Pons and Fleischmann how they measured the excess heat. The physicists were thinking that the heat commensurate with their infinitesimal neutron signal would be equally infinitesimal, and certainly unmeasurable. Fleischmann replied that they used very sensitive thermometers. And according to Decker the BYU physicists thought, "Boy, they must be *really* sensitive."

Eventually Jones told Pons and Fleischmann that he was ready to publish his data. He recalled, "I had this debate. The scientist in me says, Look, you've been working on it, publish. They don't have any neutrons. I thought, Well . . ." Jones considered this his deferral to the Golden Rule. "If they were about ready to publish," he said, "I would want them to tell me. So I told them that, after two years, we're getting ready to publish."

Fleischmann argued that Jones shouldn't go public. He was worried that if Jones published the discovery would be lost. Thousands of scientists would flood into the field. Once again the BYU physicists could make no sense of the logic. They knew what kind of neutron detectors the rest of the world had. They thought Czirr's were as good as any. "We're not worried about the rest of the world," Decker said. "We think we are as good scientists as anyone else out there. So we're not going to worry about it."

Fleischmann argued that cold fusion had weapons potential that could be dangerously destabilizing. What if the Soviets developed it first? He did, however, seem seriously concerned with the weapons question, as Ron Bullough had pointed out. (Two months later, however, Fleischmann testified before Congress that he saw no military applications for cold fusion. Perhaps he didn't want to jeopardize the Utah attempt to procure federal funding for cold fusion research.) In any case, the BYU physicists failed to see the weapons potential of a device that emitted two neutrons per hour.

Finally, Fleischmann appealed directly to Jones's professed sense of fairness. He said he and Pons had worked on cold fusion for years as well, but they needed eighteen more months. If Jones went public before then, he and BYU would receive all the credit. Jones argued that he had been invited to give a talk at the American Physical Society meeting, and he wanted to submit a paper beforehand, and Gajewski at the Department of Energy was telling him that if he wanted DOE funding on cold fusion, publishing a paper would help tremendously. He had no choice but to publish.

As an alternative, Jones suggested that Pons and Fleischmann return to

BYU Monday morning with a working cell. They would test it under Czirr's neutron detector, and, if neutrons appeared, they would write up the result jointly. Fleischmann reluctantly agreed.

Dan Decker later remarked that the BYU physicists simply had no idea what Pons and Fleischmann had observed experimentally or what the two chemists were thinking. As Decker saw it, the BYU results and the Utah results were so dramatically different that the BYU physicists could have published their work and been no threat whatsoever to Pons and Fleischmann. Very few physicists would have even bothered to read their paper. A handful might have tried to reproduce their experiment. No one would have conceived of the BYU-variety piezonuclear fusion as salvation.

By Decker's logic, Pons and Fleischmann could have then published two years later and received all the credit and acclaim they deserved. This line of reasoning illustrated just how little Decker understood, even four months after the announcement, about the ambitions of Pons and Fleischmann and the University of Utah.

Paul Palmer, by contrast, had an odd premonition during the lunch. Afterward he wrote in his log:

> 23rd of February 1989—Visit by Stanley Pons and Martin Fleischmann. U of U. This was a fun day. Pons very quiet. Fleischmann an old time con artist—maybe. At least he is so good that neither Bart, Steve nor I could tell whether or not we were conned. But we knew we'd been conned, but we didn't know how.

To Bart Czirr, Monday morning would be the ultimate test. They had known for several months that Pons and Fleischmann had tremendous electrochemical knowledge and no neutron detector. "If I could say there were no neutrons coming out," Czirr summarized, "that would be a very, very strong statement. If it melted my detector, that was also a very strong statement. . . . Either we were going to find a lot of neutrons from their big monster cell or we weren't going to find any neutrons. . . . Their whole program was going to be in big trouble if we didn't find a whole bunch of neutrons from their big cell."

The BYU physicists all had an initial prejudice. They felt that when they put Pons and Fleischmann's professional cell in front of Czirr's neutron detector it would remain as unresponsive as it had when they put their dinky, little baby-food jars in front of it. Nonetheless, they discussed what strategy they would pursue if the Utah cell did emit copious neutrons; their own data would look "piddling" in comparison. They decided that even if the neutrons emitted by Pons and Fleischmann's cell fried Czirr's detector, it would be of some historical interest to publish their own data.

All of this became irrelevant on Monday, because Pons and Fleischmann never arrived. They were expected at 8:00 in the morning. At ten Pons telephoned Czirr, who had asked Pons to convince one of the Utah physicists to help him with his neutron detector. "We expected you guys down here at eight o'clock," Czirr said.

"No, no," said Pons. "We aren't going to be down. We have a graduate student who was supposed to set the cell up so we could come down, and his father died over the weekend so he had to go to the funeral. So he won't be able to set the cell up."[7]

Pons said he'd call at the end of the week. And that was it.

When Czirr relayed the message that the two Utah chemists had stood them up, Jones and his collaborators discussed what to do next. As Palmer remembered it, Jan Rafelski, who had flown up from Arizona for the occasion, now argued that they had to publish quickly. "We were floundering around," said Palmer, "saying, 'Well, let's see what can we do to make this work better.' And Rafelski said, not quite this unkindly, 'You dumb guys don't have a clue as to what you're doing. You don't have any theories. There's no direction. You don't know what metals to use, what to do next. Listen, if you take all of the foreground and all of the background and add them up, and subtract one from the other, and the foreground signal is positive, that's enough to publish on. Let everybody else figure out what you did. Why you couldn't make it work twice the same way.' And so he said that, and we said, 'Yeah, that's right. Let's publish it.' And we decided to publish it."

Their goal, as Jones now told them, was to publish as quickly as possible. He said their funding and future research depended on it. The following day, Jones informed the DOE Office of Patent Counsel that he intended to submit patent applications on "inventions having to do with piezonuclear fusion."

During the following week, Jones and Pons spoke several times. They still discussed the possibility that the two chemists would come down with a cell; then they would write a joint paper that they could publish before the American Physical Society meeting.

On February 27 Jones noted in his logbook that he and Pons had discussed obtaining heavy water for the experiments. ("Pons: I can get clean D_2O.") And on March 2:

Pons:

    —when coming

    . . .

    —jointly publish data acq'd at BYU
       —joint paper & authorship
    -Gaj doe plan to get paper in before May mtg (APS)

But Pons and Fleischmann had little or no intention of collaborating. Pons's telephone calls to Jones appear to have been a stalling tactic while he and Fleischmann determined how to handle the situation.

The more revealing phone call was one Pons made to Doug Bennion, the electrochemist Jones had recruited earlier in February. Bennion, who had been at BYU since 1980 and had known Pons since the latter had come to Utah, recollected that the call was one of the strangest he'd ever received.

Pons wanted to know when Bennion had begun working with Jones. Although Pons didn't say this, he believed that Jones had recruited Bennion back in September, after reading Pons's proposal. In fact, Jones at that time had discussed cold fusion with another chemist, James Thorner, who had introduced him to Bennion after Gajewski told him to get an electrochemist.

Pons jumped from one question to the next. Bennion later said he felt as if he were being interrogated. Was he working for Jones? Was he building cells? Did he really expect to do electrochemistry for Jones? Yes. Yes. Yes. When Bennion finally got around to asking Pons what this was all about, Pons responded with obtuse warnings: "Be careful! You are going to get hurt; you better not do this." Bennion felt Pons was trying to frighten him off the project, but the more Pons talked, the more intrigued Bennion became. If this was a way to produce massive amounts of energy, well, then, that was exactly what he wanted to do. Bennion's one legitimate concern was that the cells would be dangerously radioactive. When Pons warned him that they could explode, Bennion asked him several times what he meant.

"Do you mean neutrons are coming off? We are going to get hurt by radiation?" Bennion asked.

No, Pons said.

So Bennion asked how Pons could possibly be getting energy without neutrons. Pons told him the meltdown story. Bennion later remarked that he worked with batteries and chemicals for a living. Explosions were an unfortunate aspect of his life. So what? Once again, he asked how they could be seeing a nuclear reaction without radiation. Pons only warned Bennion once again to be careful; then he hung up.

The outing at BYU strengthened Pons and Fleischmann's convictions on two points. It confirmed their suspicion that Jones had pirated their theory for inducing fusion. They reached this conclusion because Jones, a physicist, was indeed doing electrochemistry, not to mention doing it with palladium electrodes, just as they were. (It probably helped that Jones kept referring to his two or three years of concerted effort, while Pons and Fleischmann saw an amateurish experimental setup that couldn't have represented more than a few months of work.) Why would a physicist think of doing anything electrochemical? From a chemist's viewpoint, it looked as though Jones had thrown together an experiment that made no chemical sense and was trying to scoop them.

More important, the visit to BYU seems to have further persuaded Pons and Fleischmann that they had generated room temperature fusion. Their reasoning seemed inarguable: their cells had to be radiating neutrons because Jones's baby-food jars were radiating neutrons, and Jones was employing a technique that they had pioneered. If the electrolysis technique did not lead to nuclear fusion, why else would Jones be working so diligently to steal it? And why else would he be insisting that he had to publish it? What could be more self-evident?

University of Utah president Chase Peterson came to embrace this logic and boiled it down to its essence: "She must be a good-looking girl," he observed, "if somebody else wants to date her."

*W*hen Pons and Fleischmann returned from BYU, they reported to Jim Brophy, Utah's vice president for research, that their cold fusion work had been leaked. Brophy explained to Chase Peterson that the leak had traveled from the Department of Energy to Brigham Young University. Peterson was understandably unhappy, but he was prepared to take control. Cold fusion was no longer just a scientific issue; it had become to Peterson an administrative one as well.

Peterson had been in a bind since he became president in 1983—he had the highest aspirations for his university but limited resources. Given those circumstances, he may have been ripe for cold fusion. He'd had his eye out for salvation, and once he believed he had seen it, he wasn't about to question it, or entrust it to anybody else.

Peterson's ambitions for his school—and his state—grew out of the fact that he was the product of two cultures. On the one hand, as he noted at his inaugural convocation, he had deep Utah roots. He was born there in 1929, when his father was president of Utah State Agricultural

College, now Utah State University. Being raised on a college campus informed Peterson's life with a devotion to the educational process.

At age fifteen he won a scholarship to Middlesex School in Concord, Massachusetts. This may have been a defining experience, for he spent twenty-nine of the next thirty-four years living and working with those he would later refer to as the Eastern Elite. One of his colleagues suggested that "a scholarship student from the sticks of Utah attending such a hoity-toity institution would be affected for life."

After Middlesex, Peterson went to Harvard and Harvard Medical School. He did his internship at Yale–New Haven Medical Center and returned to Salt Lake City for all of five years in 1962 to teach and practice in a clinic. But the lure of the East won out again. Back he went to Harvard for eleven more years, first as dean of admissions, then as vice president of alumni affairs. It wasn't until 1978, at age forty-nine, that Peterson returned to Utah to become vice president for health sciences at the U.

After March 23, Peterson fell back on his belief in the Eastern Elite's superiority complex to rationalize the shellacking he received from the more prestigious scientific institutions, which he insisted were just "turf protecting." No one had data disproving Pons and Fleischmann, he would insist; it was simply a battle between "hinterland science and science on the coasts."

This delusion reveals what Peterson had in mind for cold fusion all along. He wanted to transform Utah and the university into an intellectual and economic force to be taken seriously.

Indeed, this ambition to be taken seriously seemed to be endemic to the state of Utah and fed an institutionalized insecurity complex. Rod Decker, a local political pundit, put it this way: "If you ask a lot of people in Utah what they fear, it's that everyone is going to laugh at us. We want money and we want respect. We would very much like to be respected, not as Nevada is for gambling, or Wyoming, for coal, but for brains and for talent." Thus, Peterson, two months after the announcement of cold fusion, was quoted in The Salt Lake Tribune saying that his university was in a fight against what he called the Utah effect, "which causes people to believe 'This may be important, but it may not be believable' because it came from Utah."

The state's commitment to intellect reveals itself in the fact that, per capita, Utah had the most college graduates, advanced degrees, and Ph.D.'s in the nation. But the locals still realized that, if their state were known as anything, it was as the home of the Church of Jesus Christ of Latter-day Saints, the Mormons. After cold fusion broke, for instance, the London *Sunday Telegraph* referred to Utah as "a scenic and empty state whose best-known contribution to Western culture until now has been the Mormon Tabernacle Choir."

For 150 years the local population had been living with the irksome fact that their church was viewed by the outside world as nothing more than a particularly wealthy and successful cult—and one with more than its fair share of exotic rituals, including, of course, polygamy. Lest the people of Utah be allowed to forget, Brigham Young, "the American Moses," had had twenty-four wives. Although the church had officially disavowed polygamy at the turn of the century, the practice still lived on, to the chagrin of many, in backwaters of the state. The last front-page news to come out of Utah before cold fusion was in January 1988, when the Singers, a family of radical Mormon polygamists, held off police in a bloody and lengthy shoot-out. In the words of historian Charles S. Peterson (no relationship to Chase Peterson), "Most Americans judged Mormons first unique, then foreign, then alien, then the enemy of American life. Mormonism seemed a counterculture that rejected America's progress, harked back to intolerance, mixed church and state, and threatened property."

A less specific variant of this judgment has been proposed by Paul Fussell, an Eastern Elite literary and cultural critic. In describing the class structure of America society, he wrote off Utah thus: "Indeed, it seems a general principle that no high-class person can live in any place associated with religious prophecy or miracle, like Mecca, Bethlehem, Fatima, Lourdes, or Salt Lake City."[1]

The recent history of Utah hadn't managed to ameliorate the statewide image problem, or the insecurity of the state's residents. Standing in the lineup alongside the Singers was Mark Hofmann, of the Salamander forgery trial. Hofmann's penchant for letter-bombing his associates exposed him as a forger who had sold fake Mormon documents to highly placed members of the church. Then came the story of Gary Gilmore, the local murderer who was executed in 1977 and immortalized in Norman Mailer's book *The Executioner's Song*. Salt Lake City had also earned some lasting fame as the "scam capital of the country," because it generated much more than its fair share of securities fraud. Utah seemed to be cursed as the home of some of the newspapers' most sensational crime stories.

One shining exception to this public relations nightmare was Barney Clark and his artificial heart, and that had been Chase Peterson's show. He had done a tremendous job, but he still couldn't win outside Utah. He was swinging for the fences, said his critics, trying to use Barney Clark to turn Utah into the "bionic state," the artificial organ equivalent of Silicon Valley.

Rarely, however, did they think of it that way in Utah. Peterson's handling of the Barney Clark episode increased his standing as a candidate for the presidency of the university, although he would have been a strong candidate in any case. (When a *New York Times* reporter asked Peterson what role he might tackle next, "he shrugged his shoulders and said perhaps a tour of duty as a Mormon missionary in Bolivia. Perhaps politics. In 1981, only three years after he returned to Utah, Dr. Peterson's name was mentioned in Democratic circles as a potentially attractive nominee for the United States Senate." He had proven himself to be a good administrator and a promoter of his university. He proudly referred to himself as a "booster.") He was also a Mormon, an implicit prerequisite for the job. And he was neither overly righteous about his religion nor dangerously iconoclastic.

When cold fusion came along, Peterson had been president of the university for five years, and it had not been a smooth ride. He had shown a talent for fund-raising and lobbying. He had cemented solid relationships with the governor and other members of the political leadership. But whereas his predecessor, David Gardner, who had left to take control of the University of California system, had run the U when the worst problem in the state seemed to be how to spend all the surplus revenue, Peterson was not so lucky.

The fortunes of the state of Utah tend to go up and down with the price of oil. The Utah economy had spiraled up with Gardner and had spiraled downward with Peterson. When energy prices dropped in the 1980s, Utah was hit almost as hard as Texas. The market for timber and metal mining, two usually bankable industries for the state, also deteriorated in the eighties with the rise of the dollar. Two of the largest employers in Utah, Kennecott Copper and Geneva Steel, shut down entirely for two years in the midst of Peterson's reign.[2]

Since 1981, according to *The Salt Lake Tribune,* Peterson had been forced to cut his university budget on eight occasions. In 1986, the worst year of all, Peterson had to cut $9 million and 200 jobs. Many professors had not had a raise in four years. They were earning, on average, 20 percent less at Utah than they would have gotten at comparable institutions.[3] Some of the more prestigious faculty members had moved on, including William DeVries, the surgeon who implanted Barney Clark's artificial heart. (Even after the cold fusion announcement, Peterson appeared overly anxious that he would lose his two cold fusion chemists as well. "We don't own Dr. Pons and Dr. Fleischmann," he said. "They are independent. They could go to Switzerland. They have been invited to a lot of places.")

In the fall of 1988, the year before cold fusion, Peterson barely managed to head off a statewide tax reduction initiative that would have cut his funding by another 11 percent. Locals recalled that, at one point during the year, he took to carrying out his own garbage to set a frugal example for his faculty and employees.

So Peterson was more than primed when Jim Brophy told him about cold fusion in the fall of 1988. Ed Yeates, a local television reporter, was doing a story on the university's latest financial plight when Peterson took him aside and told him that the U had something big in the works, but he couldn't say what it was. As Yeates recalled it, "He said, 'I don't know when it will happen, but it will be soon. And it involves chemistry.' " Peterson also mentioned that it could lead to Nobel Prizes. It was rumored that Peterson told the governor, Norm Bangerter, essentially the same thing in December or January. Also about this time, perhaps late December, Brophy announced to the president's cabinet that Pons and Fleischmann had come upon a new way of creating fusion.

Jim Brophy's role in cold fusion was crucial. He appeared to be incapable of doubt. "If you come down to the question of do I believe there is science there," Brophy would say, "my answer has always been yes." As the story evolved after the announcement, Brophy seemed willing to report whatever he felt was necessary to sell cold fusion or, in his words, "to put the best possible honest face" on any cold fusion–related development.[4]

Before cold fusion came along, the faculty at Utah seemed to have nothing but respect and fondness for Brophy. He was considered invariably straightforward and honest. He was known as a facilitator. In a job that was usually the provenance of administrators obsessed with proper protocol and bureaucratic consideration, Brophy could get things done.

He had come to the U from the Illinois Institute of Technology during Gardner's tenure, and Peterson had kept him on. Brophy was one of the few non-Mormons in the administration. Joe Taylor said one reason he liked Brophy was that at get-togethers at Peterson's house, "neither one of us is afraid to have a drink." Indeed, Brophy, who grew up in and around Chicago, was a childhood friend of Hugh Hefner. Hefner had been the best man at Brophy's wedding, and when cold fusion broke, Hefner wrote to Brophy, as one of Pons's chemists remembered it, "congratulations on this fusion, etc. As you are aware we've been conducting our own fusion research here in the mansion for some time." And when Hefner got married, Brophy apparently sent him a nineteen-dollar toaster and two cold fusion T-shirts, which had become hot commodities at the Utah bookstore. (One Utah administrator later suggested that cold fusion was Brophy's gift to the scientific community, the equivalent of a nineteen-dollar toaster.)

Brophy's faith in cold fusion helped convince Peterson of its potential, and Brophy gave weekly and uncritical progress reports on the research at cabinet meetings. The cabinet members were sworn to secrecy, and in the minutes the word *fusion* was never used. It was referred to as the "F-word."

Peterson's first administrative decision in the escalating cold fusion crisis was to minimize the damage done by the leak of information to BYU. He called Norm Brown, director of the U's Office of Technology Transfer, on March 2 and requested that they file a cold fusion patent as soon as possible.[5] Brown in turn called Peter Dehlinger, a patent attorney in Palo Alto, California. Dehlinger had a doctorate in biophysics from Stanford and had worked in the field for five years before picking up his law degree. The two men had worked well together in the past, and Brown wanted Dehlinger to handle the patents because he also thought the U would need the expertise of a physicist. Dehlinger readily agreed to take on the assignment, then spent the weekend in the library "trying to figure out fusion."

On Friday, March 3, Peterson phoned Jeffrey Holland, the president of BYU, to arrange a summit conference between the two universities. When he couldn't get through to Holland immediately, he went to Jae Ballif, who was provost and, coincidentally, the first cousin of Peterson's wife. The two families were very close.

As Ballif remembered the ensuing conversation, "He talked to me about the need for us to meet together and asked if I was aware of this research going on. I told him I knew a little bit about it but not much. Chase said that there was some question about provenance of ideas, and he informed me for the first time that they had people there doing something on the same subject. And he said it's of gigantic proportions in terms of its economic and scientific importance, political importance, and we needed to give it first priority and meet together."

Ballif agreed. He reached Holland and arranged the meeting for first thing Monday morning.

By all accounts, Peterson went into this meeting planning to come to a diplomatic and equitable agreement. The two schools had successfully collaborated in the past, and he had no reason to doubt that they could do so again. To Peterson, this spirit of cooperation was the Utah way;

indeed, he believed it was unique to the state. This was the flip side of the Mormon experience. The locals liked to say that they knew how to circle the wagons. This cohesiveness arose not so much out of the Mormon religion but in response to the years of persecution. The Mormons had fought the army of the federal government over the right to exist in 1857 and fought the government again forty years later over the polygamy issue before they could obtain statehood. More than anything, from the nineteenth century onward, the Mormons had fought for respect.

After setting up the BYU meeting, Peterson went to recruit Joe Taylor. As Taylor recalled it, Peterson was obviously agitated and argued that a public announcement by Steve Jones would cut the ground out from under their own effort. He said Jones was offering a collaboration, but Pons and Fleischmann would have none of it. If all parties could sit down around a table, which was what Peterson had planned, he was sure they could work out a deal. If they couldn't settle on a collaboration, they would at least agree to postpone any public announcements, then make a joint announcement when both sides were ready.

This strategy struck Taylor as a mistake, even at the time. "I didn't know why I thought it was a bad idea," he said, "except it didn't seem like the kind of thing that would lead to anything good."

Peterson wanted Taylor to accompany him to BYU as a witness to any agreements. Brophy, who would have been the natural choice, was going to be out of town on Monday, so Taylor agreed.

After Peterson left, Taylor discussed the situation with Brophy, who confirmed Peterson's account. Taylor said he thought Peterson's strategy was a bad one, and Brophy agreed.

"Do you think we should talk him out of it?" Taylor asked.

"No chance," Brophy said. "It's all set up. He won't back out now."

At BYU, Ballif remained baffled over Peterson's statement "that there was some question about provenance of ideas." Ballif went to Grant Mason, dean of the College of Physical and Mathematical Sciences, who briefed him on what he knew of the controversy between Pons and Fleischmann and Jones. The two then called in Jones, who gave his version of why Peterson might have been so anxious to hold the summit meeting. Jones assumed that Peterson was concerned about whether or not Jones had begun his cold fusion work independently.

Ballif found Jones "so believable and so open" that he decided the best way to get Peterson to understand the absurdity of any possible piracy accusations was to hear Jones himself talk about what he'd been doing.

Already Ballif and Jones were misinterpreting the reason for the summit conference. (Indeed, Peterson hadn't told Ballif exactly what he had in mind, only that they had to meet and discuss it immediately.) All the same, Ballif suggested that Jones gather the documentation of his history of cold fusion work, so he could present it Monday morning and clear the matter up.

Paul Palmer and Bart Czirr were pessimistic about the meeting. Palmer's notebook entry for March 6 reads:

> Bart said Pons and Fleischmann are going crying to the president for him to come—to bear down on us and stop us before we get beyond our discovery of the heat engine that drives plate tectonics—and may power industry. They don't want us to get the Nobel prize, when it is in their grasp.

Czirr also predicted that it would be bad business to tell Pons and Fleischmann when BYU had begun its cold fusion research, because the Utah scientists would concoct a scenario in which they had begun one or two years earlier than Jones had, which appeared afterward to be what happened.

During the forty-five-mile drive from Salt Lake City south to Provo on the morning of March 6, Pons and Fleischmann talked about their experiment. Taylor later remarked that he learned more in that drive than he had during the two months of Brophy's briefings to the president's cabinet. Pons and Fleischmann talked about the meltdown, and Taylor and Peterson were suitably impressed. Fleischmann said, as he had told Jones two weeks earlier, that they would have liked another eighteen months to pursue their research because the abundance of tritium and neutrons they had measured was not sufficient to explain the magnitude of the heat they had seen. Neither Taylor nor Peterson thought to ask how great this discrepancy was. Taylor later said that he wished he had.

The summit meeting began shortly after 9:00 A.M. in Room 301 of the BYU Administration Building. Picture windows framed a view of the mountains. On one side of a large table sat the BYU contingent: Jeffrey Holland, who was soon to leave his presidency to become one of the Seventy, which was the third level of leadership of the Mormon church; then Ballif; Lamond Tullis, who was the associate academic vice president and a Harvard-educated political scientist; and Steve Jones. Across the table, facing the windows and the mountains beyond, were Pons, Fleischmann, and Taylor.

Peterson conducted the meeting and stood through most of its three hours, often at the blackboard at the head of the table. The purpose of the meeting, he announced, centered on his hopes that the two universities could come to a managerial, social, and political, if not scientific, agreement on cold fusion. While he spoke, he chalked little boxes and arrows on the board to illustrate his points or glanced down at his scribbled notes to keep his place.

Peterson proceeded to lay out the dimensions of the cold fusion property: the potential of a limitless, clean source of energy in a world beset by environmental problems and dependent on Middle Eastern oil. He described nuclear fusion and how it differs from nuclear fission; he even sketched the two mechanisms on the blackboard. Tullis couldn't help but wonder why Peterson had gathered all this high-level administrative power only to make them sit through an elementary physics lecture. Even if they weren't scientists, they were all literate.

Peterson moved on to the huge sums of money involved in cold fusion. How could they assure that they developed this technology in Utah and didn't lose it to the technological wizards in Boston or Palo Alto or, worse, Japan? He discussed the impact the potential wealth of OPEC would have on the local and the national economy. He even broached the weapons issues. After all, cold nuclear fusion appeared to produce radioactive tritium, which is a necessary ingredient in the production of nuclear weapons.

Peterson said he believed that a satisfactory agreement between the Y and the U could be reached. He sketched out his views on how the scientific credit could be shared, hoping that his two chemists and the BYU physicist could publish simultaneous articles announcing the discovery. He hoped that they could work out the issue of patents and future research support. This was critical because Nobel Prizes were at stake, not to mention billions and billions of dollars.

To the BYU contingency, Peterson's discourse seemed to roll on interminably. Tullis's thoughts wandered over the question "How is it that Nobel Prizes are won?" Are they won by the development of a political agenda or on the basis of a substantive contribution to the science of our times? He was not so naive that he didn't know it was some of both, but . . .

Eventually Peterson brought his sprawling presentation to an end.

Ballif then said, "I think we should hear from Steve." Jones used his own hour to cover his work in cold fusion.

First, however, Jones wanted to respond to Peterson's talk of the billions of dollars to be made from cold fusion. As Jones saw it, cold fusion was only a scientific curiosity, certainly nothing that would save the world. He reached into his small blue bag and retrieved a tiny flashlight. Flipping the light on, he said in his soft voice, "Look, I don't mean to be rude, but we've been looking at this process for years now, and it's just not an energy producer. If you could ever get enough energy to light a flashlight, I'd be extremely surprised."

Fleischmann then countered that they had generated enormous amounts of heat from their fusion cells and had had an explosion. Fleischmann may have been warning Jones that they were playing with dangerous stuff, but Jones didn't buy it. This further convinced the Utah contingent that Steve Jones, physicist though he might be, was ignorant of both the science and the potential of cold fusion.

Jones's insistence that cold fusion was not worth the trouble to patent made Joe Taylor momentarily optimistic. Obviously they were talking about two kinds of phenomena. Jones must have been seeing something quite different from what Pons and Fleischmann were dealing with. Maybe he wasn't performing the experiment correctly, certainly not the way Pons and Fleischmann were.

None of the assembled administrators seemed able to grasp the fact that the effect Jones believed he had observed and the effect Pons and Fleischmann believed they had observed differed from each other in magnitude by a factor of one trillion. Jones would later take to describing the difference as though he had observed an effect comparable in power to a single dollar bill, and Pons and Fleischmann were talking about an effect comparable to the entire U.S. national debt—$1 trillion.

Then Jones began patiently to document his personal history with cold fusion. This was the same account that the scientific community would hear frequently in the next few months. Jones made a special point of displaying his lab book entry from April 7, 1986, on which, he said, he had discussed the very idea that Pons and Fleischmann had accused him of pirating. This page was notarized. It was signed by BYU's patent attorney.

Notarized?

Even Ballif, the one who had asked Jones to document his history, was stunned to see that Jones had thought to notarize his notebook. Nobody did that. Still, it seemed to make absolutely no impression on the Utah chemists.

Finally Jones came to the end of his history lesson, and Peterson got to his real agenda. He asked that Steve Jones refrain from publishing his research. He said Pons and Fleischmann needed another eighteen months to do their seminal experiment, and at that point they could all publish simultaneously.

No one apparently thought to ask Pons and Fleischmann how they could be so confident that they had discovered anything at all if they needed eighteen more months to prove it. Ballif, for one, assumed that, like all scientists, they simply would have liked time to amass additional evidence.[6]

Jones explained that he had been invited to give a talk on cold fusion at the American Physical Society's spring meeting on May 3. Peterson asked him to cancel the talk. If Jones spoke before Pons and Fleischmann were ready, and before all the possible patent issues were finalized, his talk would constitute what patent attorneys call a public disclosure. The last thing Peterson wanted to see was a premature public disclosure of cold fusion from BYU. Such an act would assure that anyone in the world could steal the Utah discovery and profit from it. Utah could end up without a dime.

But Jones refused to yield on this point.[7] He said that he was clearly ready to publish, and that the Department of Energy, which had paid for the work, was virtually forcing him to talk about it. In fact, he showed the abstract describing his talk that he had sent off to the APS offices a month earlier. Jones thought he was being open and honest, showing and telling the Utah contingent everything. The abstract said he had "accumulated considerable evidence for a new form of cold nuclear fusion."

Now Peterson seemed truly worried, which even Jones noticed.

"Are you serious that you really need to report this right now?" Peterson asked. "What if more time would allow this thing to mature more importantly, for the universities, for the state?"

Jones said he had no choice in the matter.

Peterson couldn't talk him out of it. Jones's intransigence made Pons and Fleischmann furious. Cancel an invited lecture? No, no. I couldn't possibly. The two chemists were thinking, What's so important about an invited lecture? We get requests to speak all the time.

Jones then added that he was scheduled to give a campus colloquium on cold fusion in just two days.

As Peterson recalled it, he responded, "Gosh, Steve, you understand this may have some importance, and if there's a student reporter who happens to be taking physics that day and wanders in, that can be put in the student newspaper and then it can go to the national pickup and now you've lost your patents."

Peterson believed that he was talking very openly. If he was trying to protect patents for his university and cheat the BYU physicists out of what was justly theirs, he would have been much more cunning and consciously avoided the subject. But every time the BYU group insisted that they were not interested in patents, he insisted that they ought to be. Peterson felt that the best possible outcome was for the two universities to pursue cold fusion as a joint venture: "But they were not showing interest," he noted later. "[Their position was that] this is all for the good of mankind and so forth. I wanted it to be for the good of the state of Utah."

Jones agreed to cancel the colloquium, but he would not back out of the invited talk in May.

They turned to the question of a collaboration. All the outward civility, however, could not mask the fact that Pons and Fleischmann were in no mood to have Jones in their laboratory and vice versa.

The conversation worked back around to Peterson's original idea of simultaneous publications. Pons and Fleischmann needed eighteen months, but now May 3, the date of Jones's invited talk, was the outside deadline. If the Utah chemists wanted to get the word out and share the credit, they had to do it by then. In fact, Jones's abstract would appear in print the first week of April.

Peterson finally suggested that they publish back-to-back articles in a journal, but Fleischmann said they were not sure that they could produce a definitive paper in time. Pons said nothing. He barely uttered a dozen words through the entire meeting.

Jones suggested they publish in a physics journal, in particular the most prominent, *Physical Review Letters*. Fleischmann thought this was too abstract. He said his preference would be to publish in *Nature,* a prestigious British journal. Should *Nature* not work out, he knew of an electrochemical journal that would surely accommodate them. But Jones argued that the BYU work, at least, would not be appropriate for such a narrow chemical journal, and he preferred to publish in what he later called "some reputable, refereed journal." He would not concede the point. After all, he was the one who was ready to publish.

They let it go at that. Jones would talk to the editors of *Physical Review Letters,* and maybe *Nature* as a substitute should *Physical Review Letters* be unable to publish fast enough.

Peterson then insisted that it would be unconscionable for Jones to publish or discuss the research before this back-to-back submittal. With

Jones having agreed to cancel his campus colloquium as well as a scheduled talk by one of his students, the BYU delegation thought that they had been both generous and accommodating.

Later Peterson said that his people hadn't agreed to publish simultaneous papers so much as they had acquiesced to the idea.

Joe Taylor, who had remained strictly an observer, couldn't help but think that they had spent too much time on pleasantries and never allowed their real feelings to show. Here were the presidents of two neighboring universities bending over backward to be polite to each other. It was Peterson's show, so Taylor didn't intrude, but he later wished that he had taken Peterson aside or requested a ten-minute break. If they had just gotten the chance to discuss the situation alone, he thought, they would have realized that it wouldn't work.

By contrast, Lamond Tullis, whose field of study was the politics of international narcotics trafficking and who did not appear naive, thought they had been in perfect accord. When the meeting broke, they shook hands all around. "That's part of the local custom here," Tullis explained. "[And] you don't shake hands vigorously with people with whom you still have substantial disagreements without noting that those disagreements exist."

So after three hours of near total misunderstanding, the men strolled across the BYU campus for lunch at the faculty club. (Jones excused himself, saying he had a class to teach.) As they walked, Pons mentioned the meltdown he'd accidentally induced in his laboratory. He told Tullis that a palladium cube had melted through four inches of concrete. It was one of the few things Tullis could remember Pons saying.

Over lunch the group listened to Fleischmann tell stories of his childhood. Both Tullis and Ballif found him tremendously congenial. They wondered about Pons, however, who was still so quiet. Perhaps there was no way that they could relate to Pons. Tullis later realized that Pons must have felt surrounded by a bunch of thieves and crooks.

Taylor was uncomfortable as well. He was beginning to feel like a player in some absurd farce. An appalling situation was in the works; they had come to a compromise that represented only Utah's absolute failure to negotiate. Pons and Fleischmann were obviously seething, yet they were all having a pleasant lunch and telling stories of the good old days. It was surreal.

By the time the Utah contingent walked back across campus to the parking lot, Pons and Fleischmann had begun cursing Jones: he was a scoundrel; he was trying to run off with their work.

As Pons, Fleischmann, Taylor, and Peterson drove back to Salt Lake

City, the chemists insisted that Jones's hour-long documentation of his research in cold fusion meant absolutely nothing. Peterson later recalled that one of the chemists, he didn't remember which, said, "Yes, when I was a boy, I thought of going to the moon. That doesn't mean I was doing rocket research before Goddard."

All four concurred that the newly minted agreement with BYU was unworkable. They had negotiated themselves into a no-win position. As Pons and Fleischmann saw it, any way this simultaneous submission deal played out, Jones would get more than he deserved, and they would get less. They still believed that Jones had no right to publish his electrolysis results, and now they had traded away any leverage to stop him. For instance, if Jones simply published his geophysical evidence for fusion, which was the only aspect of the business that was clearly his, then their "more cosmic and important paper was going to be given the same kind of treatment that a minor geochemical paper was given."

They would have to conceive another strategy. "I guess we have no choice," said Fleischmann repeatedly. "This is getting to be impossible to contain. We are going to have to make an announcement before too long."

Pons, Fleischmann, Taylor, and Peterson discussed the possibility that Pons present the data at an American Chemical Society meeting in Dallas, which would be the second week of April. That would give them a month to reconcile the discrepancies in their data and put together a paper.

They discussed calling BYU and telling them the deal was off, but Peterson felt that such an act could prompt Jones and BYU to go public immediately. He felt they couldn't take that chance. Peterson's attempt at diplomacy was looking increasingly dismal. As he drove the two chemists home, Peterson was thinking, What in the blazes do we do now?

The word that the summit conference had been less than a success trickled back slowly to BYU. Initially, Jeffrey Holland told Ballif, who told Jones, that they had done a good job and that "this is going to be a wonderful thing—the two universities working together."

Still, a few of the administrators remained puzzled as to why the meeting had taken place. John Lamb, the director of research administration, raised this question with Lee Phillips, the director of technology transfer. Lamb wanted to know if they had jeopardized their patent position by having Jones reveal so much about the history of his work. Phillips, likewise, was suspicious about why the Utah contingent would want to dig into their lab books. Neither was aware that it was Ballif and Jones who had been anxious to present the data from the lab books, that the Utah contingent hadn't cared if they heard it or not.

Several days after the meeting, Lamb bumped into Jim Brophy at a meeting in Salt Lake City and asked about the consensus at the U regarding cold fusion. "Jim stated point-blank," Lamb said, "that everyone was convinced that Steve Jones was stealing Pons's ideas from the proposal, and that they were pretty darned upset about it." Lamb characterized Brophy's manner as "quite exercised."

When Lamb returned, he wrote a memo to his superiors, noting that, in light of Brophy's remarks, the administration at the U might be prone to do anything. He suggested that they tread very cautiously, being careful not to reveal more than they already had about their own patent position. He also suggested that all subsequent contacts with the University of Utah be put in writing.

The administration, however, chose to take a more benign, wait-and-see attitude. Holland evidently felt that his university had made certain commitments and should follow through on them, in Lamb's words, "come hell or high water." The official position of BYU was that the meeting had been a complete success, and that *their* scientists had been the ones who had bent over backward "to accommodate the University of Utah delegation."

The official BYU history of cold fusion later enumerated the agreements reached in the meeting as follows: first, that the two sides had agreed to prepare and submit articles simultaneously to the same journal, assuring that "every effort would be made to get publication prior to the American Physical Society Meeting, even though this would be difficult in the short time available." And, second, that "no further public announcements of the results of either team's research would be made until after the papers were submitted for publication." The history made note of these two points because the University of Utah would violate them both.

In accordance with the latter requirement, Jones had agreed to cancel the department colloquium scheduled for March 8, at which he, Palmer, and Czirr had planned to talk. "They stifled the announcement of an important scientific discovery," Palmer said.

In the weeks leading up to March 23, if one chooses to believe Steve Jones, he never doubted that he was ready to publish his cold fusion work. After the fact, Jones said he could not recall trying to beat Pons and Fleischmann to publication or trying to force the issue by claiming he was ready to publish when he was not. This is possible, but the data he eventually published contradicts this account, as do his logbook entries.

As of the March 6 meeting, Jones's evidence for the existence of cold fusion was twofold: he had the geophysical evidence, which he and Palmer considered compelling but only because they were not professional geophysicists, and he had the neutron data, which may have been even less compelling.

On March 7, the day after the summit meeting, Jones noted that he had called Jan Rafelski, who had suggested that they run one more set of fusion cells, and that one such run can be significant by itself. He also suggested that they do a quick analysis, publish, and take their chances. Jones wrote, "If 4 sigma publish."

Sigma is a measure of the statistical significance of a signal, which is to say, the odds that the signal is real as opposed to some rare fluctuation of the background. It is known as the confidence level. Physicists talk about a three-sigma event or a six-sigma event, for instance. The higher the sigma, the more confidence that the effect is real. According to the statistics books, a one-sigma signal, for instance, has a 31.67 percent chance of being wrong compared with a five-sigma signal with a 0.000058 percent chance.

It should come as no surprise, however, that statistics seldom reflect real life, so they have to be adjusted downward. These probabilities assume that one understands the experiment perfectly, and knows and has accounted for every possible source of error, which is an impossibility. Thus, what is known in the business as the Vernon Hughes law of low-level statistics, named for an atomic physicist of note at Yale. His law states that despite the fact that a three-sigma effect appears to have a 99.73 percent chance of being right, it will be wrong half the time. This is real life.

Quite a few physicists would consider publishing a five-sigma event, but most such signals that make it to publication turn out to be delusions. Publishing a four-sigma event is either foolhardy or an act of desperation. One thing was certain: no one who had a four-sigma signal was ready to publish.

After March 6, when he began writing his papers, Jones did not run any cells. Between then and March 23, Jones and his friends were writing and analyzing the data they already had, which is to say attempting to prove that the meager signal was real, that the number of neutrons they had detected when the cells were running was undeniably larger than the number detected when they were not running. By the time Jones finished his paper on March 23, his neutron signal was considered believable primarily because it was infinitesimal.

Meanwhile, Jones and his colleagues investigated which journals might accommodate simultaneous submissions and rush them into publication in the same issue. Dan Decker called a friend at *Physical Review Letters* and was told that it would be difficult to publish by May 1, because the papers would have to be sent out to referees. Jones phoned Laura Garwin, an editor of *Nature* with whom he had worked a few years earlier on an article on muon-catalyzed fusion. Garwin told him that she couldn't promise anything on such short notice, "and she wasn't wild about back-to-back submissions, but she said she was willing to work with us."

"Fever-pitched" was how Marvin Hawkins described the first weeks of March in Pons's laboratory. The fusion cells required monitoring twenty-four hours a day, not for safety purposes but to record continuously the voltages and currents going into the cells and the heat coming out. When Hawkins's stamina ran out, Fleischmann or Pons would come in and take their shifts.[8]

Fleischmann had said again and again that they needed eighteen months, at least, to finish the necessary experiments. They discussed the time constraint frequently, asking each other the same questions: "What are we going to do? If we have to press for publication, what matrix of data can we collect to make a compelling argument for fusion?"

Hawkins remembers that they ran the calorimetry experiments until they were convinced that the cells were producing too much heat to be explained by any chemical means. Then they would take the cells apart and rebuild them and run them again until they gave the anomalous heat or burned out. Some ran for as little as a few hours before they died, and others that had been running since Christmas continued to generate more heat than the conventional wisdom dictated was possible. The data, Hawkins said, "were coming off quite reasonably. There were times when I built cells and had seven out of ten work."

Still those cells obstinately refused to emit neutrons. Fleischmann had already called Dave Williams at Harwell and explained the problem to him. The neutron detector that Fleischmann had purchased from Harwell, as Williams recalled, "should've gone berserk, and they weren't seeing anything on it."

Fleischmann had also asked Williams if Harwell was prepared to put together some fusion cells and count neutrons for them. Williams went to Ron Bullough, the lab's research director, with the request. "Just do it," Bullough said.

On March 9, Fleischmann faxed Williams instructions to make two of their fusion cells. One was the cell design that Hawkins said he had conceived in November. In his note, Fleischmann referred to it once as the "awful cell" and then wrote, "Very bad geometry for counting?" So Fleischmann apparently believed, or at least hoped, that fusion was occurring and neutrons were being emitted but that somehow the configuration of the cell prohibited their appearance or detection. Fleischmann also included much less detailed instructions for building what he called "our No. 2 monster." And here Fleischmann noted, "Much better geometry?" Perhaps the monster had seemed less reluctant to emit neutrons.

At Harwell, Williams called David Findlay, a Scottish physicist who had been recruited to help with the neutron detection. Findlay said he had this "tremendously sensitive neutron detector" up and running, and he was ready for cells. He was not optimistic. "I didn't believe it at all," he later recalled. "Somehow the idea that you get a beaker of heavy water and a few metal electrodes, connect a battery, and, bingo, neutrons come flashing out. . . . I mean, people have been trying to produce neutrons for a long time." Findlay's partner, Martin Sené, said that when he saw the fusion cells his first reaction was "hysterical laughter."

Despite the physicists' skepticism, Williams had the Harwell technicians build fusion cells to match Fleischmann's drawings and took them over to Findlay's neutron counter, only to find that they didn't fit. "There was some grinding of teeth over that," Williams said. After alterations the fusion cells were inserted gently into Findlay's detector and set to charging. It was the beginning of an exasperating wait for neutrons at Harwell.

Pons, meanwhile, was procuring physics expertise closer to home. He called the Nuclear Safeguards Division at the Los Alamos National Laboratory and spoke to a division director, George Eccleston. Pons asked what kind of neutron detection capability they had. Was it portable? Specifically, could it be transported up to his laboratory? The answer was yes. But Eccleston also said that he'd need to know about source strengths and setups, and it would take longer than Pons expected. "You can't go in overnight," Eccleston said, "and do a spectrum."

Pons also called Robert Hoffman at the radiation safety office on campus. Hoffman had a master's degree in medical physics and had some experience with radiation detectors, but he was not remotely a research physicist. However, Pons trusted him, which may have been more important. Pons told Hoffman that they needed help in observing what Hoffman called "gamma photons." Pons appears to have reasoned that any neutrons emitted by the palladium electrode might be interacting with the hydrogen in the water bath that surrounded the cell and releasing a gamma ray. In fact, the neutrons *should* have been interacting in just such a manner. This was a well-established nuclear reaction. If Hoffman's equipment could detect these gamma rays, which always have an energy of 2.22 MeV, it would be as good as seeing the neutrons themselves. And if all the neutrons were disappearing into the water bath, that might explain why Pons and Fleischmann hadn't seen any.

Hoffman was suffering from the flu during the week of March 6, but he offered to do what he could. He carried a portable detector down to Pons's laboratory and set it up directly over the working cells. The detector was a standard health physics device, known as a sodium iodide detector, and Hoffman connected it to a series of electronic devices—a photomultiplier tube, a preamplifier, and a device called a pulse height analyzer. This equipment would not only flag any incoming gamma rays but also measure the energy of the gamma ray. If the fusion cells were emitting an excessive number of gamma rays at that one signature energy, 2.2 MeV, it would suggest that neutrons had been emitted from the cell.

Hoffman let the detector run for forty-eight hours directly over the fusion cells. Then he measured the background radiation by running the detector for another forty-eight hours over a laboratory sink on the far side of the room. When he was done, he compiled the raw data, which were on pages of computer printout, row after row of numbers corresponding to gamma rays and energies, and handed them over to Pons.[9]

On March 10, Pons and Fleischmann initiated the process of betrayal. As told by all parties, their actual strategy was not premeditated; however, that it would involve a betrayal was.

That day Pons received a call from Ron Fawcett, an electrochemist at UC Davis who was the American editor of the *Journal of Electroanalytical Chemistry*. Fleischmann had suggested *JEAC* as his second choice for publication at the BYU summit conference. They had no guarantee that *Nature,* their first choice, could accommodate their request for an immediate back-to-back publication, but *JEAC* was virtually family. Roger Parsons, the managing editor, was a professor at Southampton and had been a close friend of Fleischmann's since the late 1940s. Fawcett had done his doctorate in Parson's laboratory at Southampton. Both Pons and Fleischmann had published frequently in *JEAC*.

Fawcett called Pons, as he later told reporters, simply to discuss the performance of one of Pons's students, who had been out to UC Davis on a job interview. If so, it was a fortuitous coincidence. Pons took the opportunity to tell Fawcett about the fusion project. He said that they had detected tritium, which he considered proof enough that they were on the right track. He even told Fawcett about the BYU complications, and his concerns about getting into print as soon as possible. As Fawcett told it, Pons "wanted to get something together real quick" and send it to Fawcett by overnight delivery that evening.

If Fawcett received the paper the next day, he could stamp it with a March 11 reception date, which could be important in any future priority fights. Fawcett could then route it on to Parsons at Southampton, which is exactly what he did.

Fawcett thought cold fusion sounded like a brilliant idea. Typical of Pons and Fleischmann's genius to come up with something so simple. Why hadn't anyone else thought of it?

Jones and the BYU contingent later viewed the *JEAC* paper as an out-and-out attempt to shaft BYU and break the universities' agreement. It's also possible that it represented a fail safe for Pons and Fleischmann. Publishing in *JEAC* certainly violated the agreement with BYU, but it ensured that they wouldn't be victimized by fate any more than they already had been. Pons and Fleischmann feared that if they submitted to a refereed journal, as Jones was insisting, one paper—the BYU paper, for instance—could be accepted and the other—theirs—rejected. This was one of the many worst-case scenarios of the back-to-back submission agreement.

Their anxiety was understandable. How could Pons and Fleischmann write a paper in the short time available, with the paucity of data they had,[10] that would meet the standards of the referees of *Physical Review Letters,* all of whom, coincidentally, might be physicists?

On March 13, an obviously frustrated Fleischmann sent a fax to Dave Williams at Harwell. Fleischmann explained their situation and included a provisional tabulation of the heat data they'd taken so far. He declared that this "information is still very incomplete so you can imagine how annoyed we are to be rushed into premature publication."

Fleischmann added that the rate of tritium generation in their cells was much less than it should be, considering the copious heat production, and neutron production was lower still. "What on earth is going on?"

he wrote. All of this should explain, he said, "why I am so nervous." He closed, "yours exhaustedly."[11]

That same day Fleischmann shipped two complete fusion cells to Harwell to run for neutrons. The following day Williams received the two cells and relayed a succinct message to Fleischmann: "Received cells, all well." This was not exactly the case. Finlay and Sené, who were operating the neutron detectors, hadn't observed any neutrons from the cells that had now been charging for three days.

Also on March 13, Ryszard Gajewski at the Department of Energy called Steve Jones and told him that he had spoken with Stan Pons. Pons apparently said he was still committed to publishing back-to-back articles. But Pons was also still insisting that Jones's work, as Jones wrote it in his logbook, "follows from his proposal."

That evening Pons and Gajewski spoke again, and once again Gajewski relayed the conversation to Jones. This was the same story. Pons was troubled by the fact that Jones would report electrochemical data in his paper, and he repeated the implication that Jones's work was "steered" by the proposal.

As Jones recalled it, "Gajewski was concerned about this. He really was. Gajewski is very upright. Integrity is the first thing. Gajewski challenged Pons. You want an investigation, fine, we'll have an investigation. Gajewski was talking to me: 'What do we do? I've been funding you; I need to know about these allegations.' "

Jones, too, was upset and considered himself very upright. He wanted immediate action. He disliked the idea that anyone was spreading these accusations, and he disliked the feeling it engendered in him: "I just start feeling jittery," he said, "like hot pins and needles inside. I don't like it. I mean, here's a guy spreading rumors behind my back. I called Stan and I said, 'Look, I hear again this rumor. I thought we worked through this. I showed you my history. What's going on? And why are you doing it behind my back, and why aren't you asking me if you have questions? Why didn't you raise the questions at the meeting? I'm sorry, but I'm feeling very upset.' And Stan said, 'Well, I'm feeling very upset.' And he hung up on me."

This was only the beginning. Pons and Fleischmann then called Jones back. "They had either two extensions or a speakerphone," Jones recalled, "and Fleischmann said, 'Here's a warning: you could get heat to the degree that things could be dangerous,' and he told me certain conditions should be avoided. He said, 'Look, be careful down there because you could get some real problems.' " Jones didn't know whether this was a threat or a warning.

In any case, both Pons and Fleischmann now seemed as angry at Gajewski as they were at Jones. They had yet to receive any funding from Gajewski, but they had heard that the DOE had finally approved their proposal on March 2. The two chemists believed that Gajewski was holding the money until the situation between Utah and BYU could be resolved, which meant, as far as they were concerned, that they were being forced into this collaborative arrangement.[12]

All Jones recalled was that the money had been approved but that Pons and Fleischmann didn't have it. "They were peeved about that," Jones said. "All of this was somehow mixed in there. So I said, 'Look, guys, all I'm asking is don't spread insinuations behind my back. If you have any questions ask me.' And Fleischmann agreed."

Jones then told Jae Ballif what had been passing between the two universities and the DOE. Ballif apparently suggested that the University of Utah should be asked to make a public retraction. And Jones addressed a letter to Ballif outlining his position:

> . . . I do not believe that I have incorporated any of their original ideas into my research. Rather, I have invited Pons and Fleischman [sic] to make use of our neutron detection equipment, developed at BYU over the past few years for this and other research, in conjunction with a joint research effort. Their methods are complementary to ours and there is therefore good reason to join forces, especially since cold nuclear fusion has potential (though with low probability in my mind) to be of great benefit to mankind.
>
> The charges of improprieties are serious and I wish to demonstrate that they are unfounded to an impartial review board, to put an end to these unsubstantiated rumors, and, hopefully, to redeem a severely strained agreement to perform collaborative research on cold nuclear fusion. Therefore I propose that a thorough investigation be conducted. . . .[13]

On the night of March 15, Pons enlisted legal help of a sort. C. Gary Triggs, his lawyer and childhood friend, had come to town for a strategy meeting that Chase Peterson had called for March 16. Triggs was a bulldog of a small-town defense attorney, fast talking and full of southern homilies. As Peter Dehlinger put it, you could imagine him in court charming the socks off of juries with his yarns and his mellow accent. More than anyone, maybe even Fleischmann, Pons now turned to Triggs for advice.

Once again, Gajewski relayed to Jones the details of a caustic conversation with Pons and Fleischmann. As Jones noted it in his log, Pons went into a "tirade" and read a prepared statement, alleging, once again, that there had been improprieties in the reviewing process and that Jones had taken information from the Utah proposal and used it in his research. Pons said they would not retract any accusations they had made, and he would not allow himself to be coerced by Jones or anyone else. Gajewski warned that if it came to an investigation, both universities "would be dragged through muck."

The way Pons and Triggs told it the next day, they had let Gajewski have it with "all their legal guns blazing." The two said that Gajewski had been making veiled threats, to the effect that Pons would never get a DOE grant again if he didn't shape up. So Triggs, who was apparently listening in on the extension (or speakerphone), signaled Pons, who told Gajewski he would call him back. Then Pons and Triggs hung up and discussed the situation. Triggs wrote up the statement, and Pons called Gajewski back and, with a recorder taping the conversation, read it to him.

According to Hugo Rossi, dean of the College of Arts and Sciences, Pons later said that Gajewski made incriminating statements on that tape. But another witness to this story insisted that "Gajewski smelled a rat and wouldn't say a thing."

I seen my opportunities, and I took 'em.
GEORGE WASHINGTON PLUNKITT
of Tammany Hall

*I*f cold fusion had a patron saint, that dubious honor would probably go to Blaise Pascal, the renowned seventeenth-century physicist, mathematician, and philosopher. Pascal renounced a life of science for one of faith, which many of the proponents of cold fusion seem to have done, and he wrote down the terms of the wager that, for him, made this choice inevitable. Pascal argued that to bet on the existence of God and to be wrong is to lose little or nothing. To wager correctly that there is a God is to be rewarded with an "infinity of infinitely happy life." "Let us assess the two cases," he wrote: "if you win you win everything, if you lose you lose nothing. Do not hesitate then; wager that he does exist."

Throughout the cold fusion episode, the proponents of cold fusion would subscribe to the logic of Pascal's wager. To bet that cold fusion existed and to win was to be rewarded with a payoff that, while not literally infinite, certainly seemed like it at times. To bet wrongly cost relatively nothing: a few million dollars, a few months of work, or a reputation would always seem inconsequential in comparison to the potential reward.[1]

One year later, for instance, Chase Peterson insisted that he had never believed that cold fusion necessarily was real, but that what was important was that it could have been real. Here was Pascal's wager. Peterson said, "You get burned if cold fusion doesn't work, but you sure get burned if you don't do anything about it and it does work. So you've just got to be smart."

After March 23, the conventional wisdom was that somehow the lawyers had forced the press conference on Pons and Fleischmann. Nobody knew exactly which lawyers, maybe the patent lawyers, but that was not the point. What was certain was that the two chemists would not willingly have initiated such a piece of shameless grandstanding, and the administration of a respectable university surely could not have been responsible for it. So lawyers were the natural scapegoats. But in the final analysis they were only accountable for one third of the responsibility.

When Peterson realized that the equation he had worked out with BYU was not a workable one and that cold fusion, with its prodigious potential, could not be left in the hands of the principal investigators alone, he repeated the procedure he had employed with the artificial heart six and a half years earlier.

At that time, Peterson had gathered all the key players: the surgeon, Bill DeVries, and his assistant; the regular and intensive-care nurses; the research team; the lawyers; the hospital security people; the public relations staff; and the hospital administrators who would finance the procedure. Then he said, "I'm not in charge of this, but I'm going to moderate. You describe to me what you think your role will be if we ever choose a patient for the artificial heart." They went around the room, and everyone spoke honestly and directly. Once they had worked out all the complications to everyone's satisfaction—and once they had located a viable patient, the redoubtable Barney Clark—they went for it.

With cold fusion, Peterson had a considerably smaller team to assemble. He brought in the lawyers: Peter Dehlinger from Palo Alto and two whom Pons had requested personally, C. Gary Triggs and Gary Sawyer, another North Carolina patent expert who had worked with both Pons and Triggs in the past. From the university were Jim Brophy, the vice president for research; Norm Brown, head of the Office of Technology Transfer; and, of course, Pons and Fleischmann.

This meeting convened in Peterson's office on March 16, and the lawyers began with their agenda. As Peterson recalled, "They began to say, 'This is crazy. This has got to be identified now as the Pons-Fleischmann phenomenon. It's got to be so identified. And the patent primacy rests partly on lab books and all the rest just on being ahead of the game.' "

Pons and Fleischmann demonstrated to the lawyers that they could document their provenance, which was consideration number one. They discussed dates of invention and decided, on whatever evidence, which nobody would later admit to actually having seen, that they had priority.

Peterson then shifted the discussion to the issue of economic development. He was out to assure that any economic benefits accrued by the technology of cold fusion would go to the state and the university. He had already arranged with Pons and Fleischmann that royalties would be split, one third to the two chemists, one third to the university, and one third, up to a preset limit, to the chemistry department. Now he wanted to assure that there would be royalties, that if cold fusion blossomed into a viable energy source, it would happen in Utah and not Silicon Valley or Japan.

Pons and Fleischmann now explained the status of the BYU cold fusion research. Upon reading their proposal, Steve Jones had obviously made a major shift in the emphasis of his research. They believed that his electrolysis experiments from 1986 were just one of many approaches he had tried, and that they hadn't panned out. They considered meaningless Jones's notarized lab book page—the one on which he had set down his ideas of cold fusion. They were idle scribblings. Yes, Jones had mentioned palladium, but not in any specific context, and he had never gotten around to using palladium electrodes, or at least not until after reading their proposal, which coincided with his return to the electrolysis experiments.

The two chemists said they believed that Ryszard Gajewski was responsible for initiating the dispute, and thus their present troubles. The way they had come to see it, Gajewski had been funding Jones to the tune of a few million dollars over six years and had nothing to show for it. Suddenly they appeared with a startling idea of great promise, one that would certainly justify substantial federal funding if it panned out. And, equally suddenly, Gajewski began to stall. Their funding seemed to be contingent on a collaboration with Jones. They believed that Gajewski was trying to validate somehow all that muon-catalyzed fusion money by linking it to their cold fusion. They had used their connections at the Department of Energy to check Jones's funding history, and what they had found was all muon-catalyzed fusion. They had looked up his various publications and found only that one insignificant paper on piezonuclear fusion.

Pons and Fleischmann's most important conclusion was that, like Peterson, they could not trust Jones's "naïveté." ("Naïveté," Peterson later remarked, "is one way of saying it. There are other ways of saying it.")

Peterson's congregation then decided that they owed Jones and BYU nothing. They believed that they had developed their technology independent of Jones, and they could let history, the peer review process, or the patent board determine whether Jones had taken anything from them.

Norm Brown described it as a case of apples and oranges. "We could let Jones do his thing," Brown said, "and we would do ours, as long as we did ours first. Then Jones could say, 'Yeah, me too.' Or he could say, 'Well, I found out this other interesting thing,' which is what he ended up doing. It could hurt us, on the other hand, if we say, 'Yes, Jones invented the same thing,' which is what is implied by having back-to-back publications. So we ended up thinking that we would be legitimizing Jones, building a mountain out of a molehill, if we did that."

Peterson apparently did suggest momentarily that he call Jeffrey Holland at BYU just to inform him that they were not going to abide by their agreement, but he was quickly convinced that that was not a wise idea. If he gave Holland five days' notice, or even two days', BYU might go public immediately. That was a risk they wouldn't take, so there would be no communication with BYU.

As long as the Utah patents were on file—Dehlinger had filed one on March 13 and would have a second filed by the twenty-first—they would retain domestic rights to cold fusion even if Jones announced first.

In the United States, patent rights go to the first to invent, which is to say the person who can establish the earliest date for a working prototype of the device in question. In the event of a public disclosure, anyone believing he or she has a claim to the patent has one year from that moment to file. In all foreign nations, however, with the exception of Australia and Canada, the patent goes to the first to file. Once the invention is disclosed publicly, no patents will be awarded. Thus, any hold on the foreign rights would be lost the instant Jones made his disclosure.

Public disclosure has a relatively wide definition: writing up the invention in a company newsletter, for instance; discussing the technology with anyone not party to a secrecy agreement; or releasing any information that would allow an expert to reproduce the invention might constitute public disclosure.

The abstract that Jones had submitted to the American Physical Society, even with two meager sentences on cold fusion, might constitute a public disclosure once it appeared in print. It was scheduled for the first week of April. That became the deadline for whatever had to be done, which gave the U roughly three weeks. The option of having Pons unveil cold fusion at the American Chemical Society meeting in the second week of April was no longer viable. Jones's APS abstract would appear a week before the meeting.

Peterson, Pons, and Fleischmann had bandied about the idea of throwing a press conference since the BYU summit meeting. The two chemists had already discussed the situation with Roger Parsons and Ron Fawcett, the editors of the *Journal of Electroanalytical Chemistry,* and had received their assurance that a press conference would not jeopardize the publication of their paper. (In fact, *JEAC*'s decision to publish was somehow treated as justification for the announcement. That the paper would not be peer-reviewed, but rather accepted on the strength of Parsons's judgment alone, was considered a trivial technicality.)[2]

Initially it was Peterson who suggested the public announcement, but the three lawyers apparently embraced its wisdom. Dehlinger later supposed that the decision might have gone the same way even if everyone but Peterson had been against a press conference. But such was not the case. "The fact is," Dehlinger said, "the three lawyers were arguing that there is no second place in this kind of business. Either you're there first or no one remembers you. Gary Triggs was particularly adamant because he is such a fanatic believer in Stan that he was the most upset at the idea of sharing this with Steve Jones."

Dehlinger concluded, "We all, I think, supported and were trying to overcome the resistance of Stan and Martin."

Pons seemed most concerned with the logistics problem at hand. How would they write the *Nature* paper, finish the *JEAC* paper, prepare a press release, and complete the necessary research, all in one week? But he agreed to go along with a public announcement. He trusted Triggs, perhaps even as much as he trusted Fleischmann. As Dehlinger recalled it, "Gary Triggs never doubted the science. And with that position he also never doubted that this was one of the great scientific breakthroughs of the twentieth century, and his friend Pons was going to get credit only if he was bold, and he was certainly pushing Pons to be as bold as possible." Norm Brown later observed that the relationship between

Pons and Triggs transcended that of lawyer and client. "Stan relied on Triggs a lot," Brown said. "Pons looks to him for advice, and so, de facto, if Pons wasn't satisfied, the university had a problem, and if Triggs wasn't satisfied, Pons wasn't satisfied."

Dehlinger's recollection of the meeting also had Fleischmann "almost in tears" as the consensus finally emerged that they would call a press conference. This contradicts Peterson, who later said he would never have gone on with the announcement had Fleischmann been so noticeably against it. Maybe so.

Either way, Fleischmann certainly was the most prescient about the ugliness of the deluge that would follow a news conference. And afterward it was Fleischmann who would lay the entire responsibility for the decision and the subsequent circus on the U administration. "That was the decision of the university," he said. "You can read into that anything you want." Nonetheless, at that point, Pons and Fleischmann still could have put a stop to the affair. They did not.

The administrators and lawyers had only to worry about whether Pons and Fleischmann were correct in their interpretation of their experiment. Peterson later said that he believed in cold fusion, to the extent that he did, because Pons and Fleischmann were "very competent electrochemists. They say that they've got something that cannot be explained by a chemical reaction. Well, these guys ought to know what a chemical reaction is." He also believed because Jim Brophy believed it unconditionally, and Brophy had been trained as a physicist. And Peterson was impressed with Fleischmann's brilliance, as was everyone else. Although Fleischmann was not a physicist, he appeared to have a deep understanding of the field.

It was the account of the explosion, however, the famous meltdown, more than anything that convinced the lawyers and administrators. All the other evidence of nuclear fusion—heat, neutrons, gamma rays, and tritium—paled next to the tangibility of the explosion. "Frankly," said Dehlinger, the biophysicist turned patent attorney, "if there was one thing that made me feel good and think that there is something there, it was the explosion. Yes, it can be chemical, but that didn't seem likely. Something pretty significant in terms of heat generation must have been happening. It was the anchor which many of us were using. Whenever you're doing science, lots of things can go wrong and get in the way of reproducibility. We took comfort in the few events which seemed spectacular."

In a sense, Peterson and his congregation ultimately believed in cold fusion because it was too big to question. It was Pascal's wager. "If there is any merit to this," Peterson said later, "can we afford to let it go the way of classical, normative science, that is, a very dignified, reserved announcement, a very slow and methodical process of development and substantiation?" The answer was no.

One final consideration still might have prevented a press conference, however. Brown suggested that before going public they give serious thought to the political implications of their announcement. Pons and Fleischmann, after all, were claiming that it would be possible to build nuclear weapons with the technology. Brown observed that should this be the case, as with Pandora, there'd be no going back once they opened the box. "What if this gives Qaddafi the ability to make a nuclear bomb for fifty dollars?" asked Brown. "Is this something that we really want to do without thinking about it ahead of time?"

This was somehow too profound to contemplate in the short time they had, so it was ignored. Brown recalled that Peterson and the others gave it a few seconds' thought and in effect said, "Okay, now we've thought about it, let's throw a press conference."

As the meeting concluded, Pam Fogle, the university's news director, was called in to discuss how to handle the press. She was informed that her office could interview Pons and Fleischmann the following day, which was Friday the seventeenth; then she would have the weekend to write a press release. It would have to be ready by Monday so that the scientists, lawyers, and administrators could review the draft. Be careful, Fogle was told, that no drafts of the press release fall "into the wrong hands." It sounded like a spy novel.

Fogle said the determination was made then to schedule the public announcement tentatively for March 23, one week later. Once they had committed themselves, secrecy became a greater concern than the validity of the science. Peterson, for instance, chose not to consult any of the physicists on campus. He feared they might leak news of the "F-word" to the outside world, or even leak word of the press conference to Jones. He also believed his chemistry department was the "more prestigious department." Thus, employing logic that seems dangerously specious, Peterson concluded that his prestigious chemists did not need assistance from physicists to solve a physics problem.

Peterson did seek advice from Hans Bethe, one of the legendary figures in nuclear physics. Bethe, who was eighty-two years old, still taught quantum physics at Cornell. Physicists liked to say that taking quantum mechanics from Bethe was like taking Russian literature from Tolstoy. (Peterson and Bethe were distantly related through their children's marriages. Bethe's daughter-in-law had a brother who was married to Peterson's daughter.) Peterson phoned Bethe and told him that Pons and Fleischmann had observed electrochemically induced cold fusion.

Bethe replied that this sounded very unlikely.

Then Peterson said that physicists at Brigham Young University were also claiming cold fusion and were going to publish a paper. He didn't want his university left out.

"Let BYU publish alone," Bethe said. "Let them make fools of themselves."

Peterson ignored the advice, apparently because it was not what he wanted to hear. And Bethe wasn't swayed by the persuasiveness of Pascal's wager. Indeed, as far as the wager went, to sit quietly and let Jones and BYU publicly announce the discovery of cold fusion was, in effect, to bet that cold fusion did not exist. Bethe, with his deep understanding of canonical nuclear physics, might be able to do that. Peterson could not.[3]

Peterson spent March 20 in Washington on business unrelated to cold fusion. He flew back to Utah the next day. On the flight were two Utah physicists, Pierre Sokolsky and Michael Salamon, who had been visiting their benefactors at the Department of Energy. Half of the remaining passengers, or so it seemed, were adolescent girls returning from a Washington field trip. Sokolsky and Salamon escaped their chatter by hiding in the galley.

Peterson also fled to the galley. After some brief small talk, Peterson told the two physicists that he'd been doing some reading, and he'd like to know what they could tell him about muon-catalyzed physics. Salamon and Sokolsky were astonished. Muon-catalyzed fusion was esoteric even within the refined field of nuclear physics. It was certainly not a subject that they would have expected to pique the interest of a man of Chase Peterson's medical and administrative background. In fact, the two physicists had never given it much thought themselves, although as physicists they knew of it and could explain it, which they did.

For the better part of an hour, Peterson interrogated the two on various fusion mechanisms, using correctly all the technical jargon. Later Salamon and Sokolsky recalled that Peterson never explained the motivation for his curiosity. Nonetheless, their opinion of Peterson, which had not been high, skyrocketed.

On March 21, Dave Williams at Harwell faxed a succinct message to Martin Fleischmann in Utah:

5 days now & no neutrons. Any suggestions?

Fleischmann called Williams and told him they could not sit on cold fusion much longer. Now, however, Fleischmann was optimistic. He said he was confident that he had counted neutrons, having detected apparently three times as many coming from the vicinity of the fusion cells as from the background. Although this was not the billionfold or trillionfold increase predicted by nuclear physics, it sounded convincing to Williams. (But Williams later admitted, "I don't know anything about neutron counting.")

Fleischmann also said that they had been detecting gamma rays from a cell with an eight-millimeter palladium electrode but that these rays had vanished. They did, however, seem to be registering a very definite gamma ray signal from a cell with a four-millimeter electrode. "And so he concluded," Williams recalled, "that the eight-millimeter rod had died. Therefore there was a possibility that the experiment could die, or maybe not even work. So he knew at that time that the thing was irreproducible."

The effect may simply have required a fine tuning of the cells that was hit-or-miss. Some cells worked and some didn't. Williams had his five-day-old cells removed from the neutron counter at Harwell and three new cells inserted.

That same day, Pons telephoned Steve Jones, wanting to know if he still planned to publish his paper.[4] Jones said it was almost written, and they were committed because of the APS meeting. The two then agreed that they would rendezvous to send off the papers at the Federal Express office at the Salt Lake City Airport, at 2:00 P.M. on March 24. Fleischmann was returning to England that day, and it would also be the day after the Utah press conference.

Jones noted in his lab book that when he asked Pons if the Utah cold fusion paper might be submitted early, "e.g. by Fleischman [sic], he said no." So Jones was suspicious. He simply had not asked the right question.

On the morning of March 22, Pam Fogle began alerting the local and national press to the coming news conference. Once she called Jerry Bishop of The Wall Street Journal, the story began to leak, as the administration had feared. Bishop was a ponytailed Texan whose working wardrobe ran to cowboy boots and blue jeans and who did not look like a reporter for the country's preeminent conservative business paper. But he had been a science reporter for three decades. When Fogle told him that Utah had sustained a nuclear fusion reaction at room temperature for one hundred days, Bishop refused to believe her.

After talking to Fogle, Bishop went to lunch with a free-lance science writer who was connected with the Council for the Advancement of Science Writing. Bishop described the call, and the friend reminded him of a recent meeting at which a physicist from Utah had discussed some kind of strange fusion. Bishop went back to his office, checked through his files, and found a copy of a Steve Jones paper on muon-catalyzed fusion. Bishop called BYU, managed to reach Paul Palmer, and asked what the story was with the University of Utah announcement. "All we can say is our results don't confirm theirs," Palmer told him.

From that moment, Bishop knew he had a story. His story in the *Journal* the next morning reported that a new fusion breakthrough would be unveiled and speculated that it would show "that hydrogen atoms can be forced to fuse together inside of a solid material." Bishop reviewed Jones's previous work in muon-catalyzed fusion and noted that both the University of Utah and BYU had simultaneously submitted papers to *Nature.**

As it turned out, the *Financial Times* of London beat Bishop to the story. It seems Fleischmann had gone to an old friend at Southampton, Richard Cookson, a retired chemistry professor, to get advice on giving the story to the British press. Cookson put Fleischmann in touch with his son Clive, who was a reporter at the *Financial Times*.

Young Cookson had explained to Fleischmann that the *Financial Times* would not be published on March 24 because it was Good Friday. If Fleischmann wanted the British press to get the story before the long weekend, he would have to let them publish it on Thursday, the day of the press conference. Fleischmann apparently talked it over with Pons and maybe Brophy, then gave his permission and the necessary information.

Harwell physicist Ron Bullough suggested that Fleischmann's act was motivated by national concern. "Since he's a true Brit," said Bullough, "he felt he had to do that. I think he was using Cookson to warn the establishment and the government that all hell was about to break loose."[5]

With Cookson and Bishop working the story on Wednesday morning, the news immediately got back to BYU. Like all rumors, the story appears to have evolved spontaneously as it spread. Bishop apparently

---

*Actually, that had not yet happened and was looking less and less likely.

tracked Jones back to Gajewski, whom he called to find out what he knew about the Utah press conference. That prompted Gajewski to call Jones and tell him that "all hell has broken loose" at the Department of Energy.

As Gajewski heard it, apparently misunderstanding Bishop, the University of Utah had a press release that claimed heat production by cold fusion, while simultaneously claiming that a reviewer of the proposal confirmed the result. Jones, of course, was outraged. "Baloney!" he scrawled in his lab book:

> And why should they announce our unpublished results to the press? Press release also flies in face of our agreements not to speak of results publicly until papers back-to-back were in (Friday 3/24/89).

Before the day was out, Jones also received calls from the *Financial Times,* to confirm that he had confirmed Utah, and from a broker in Boston, who apparently wanted to know if they really used palladium, in which case should he buy futures?

As the reports of BYU's confirmation escalated, Grant Mason, dean of the College of Physical and Mathematical Sciences at BYU left a message for Paul Richards, the university's public communications director:

> Apparently the U of U has released a story to the press indicating that BYU backs up or supports some research they are doing on cold fusion. [Mason] is concerned that this could be quite sensational since we do not in fact support their conclusions. We need to know if the press release went out and exactly what it said, and he also wants to prepare a statement as a disclaimer.

Richards called Ray Haeckel, his opposite number at the University of Utah, and Haeckel read him a copy of the press release, which made no mention of any confirmation whatsoever. "I can't imagine how you have this," Haeckel said, "because we haven't even sent it out yet. We're not going to send it out until tonight."

Richards reported back to Mason, who promptly phoned Jim Brophy. Brophy also insisted that Utah hadn't issued any press release, which was still true. They had been working on a draft, Brophy said, and they had been very careful not to mention BYU at all.

"But people are quoting us," Mason said. "They've got something."

"If you find out where it came from," said Brophy, "let me know. Because I don't know where."

Mason then warned Brophy that should Utah hold a press conference, the BYU administration would consider it "a stab in the back."

The phone calls continued through the afternoon. Steve Jones called Joe Ballif with the news that Pons and Fleischmann might be holding a press conference. Ballif refused to believe it: a Utah press conference constituted betrayal of a magnitude he simply could not accept. He had Chase Peterson pulled out of a meeting to take his call. He said that he'd heard a rumor of a press conference, which must not be true. And Peterson replied, "We have talked about that, yes. We intended to make some kind of an announcement." Peterson went on to say that he'd never been satisfied with the outcome of the meeting in Provo. He said he'd call Ballif back, which he never did.

The announcement was made the following day in the main foyer of the Henry B. Eyring Chemistry Building on the Utah campus. The available seating space was quickly taken by press, administrators, students, and curious scientists.

Marvin Hawkins attended the press conference with his wife, his father, and his in-laws, and managed to find seats three rows back. Fleischmann had warned him that their lives would never be the same after they went public and that the activity unleashed would be astounding. Hawkins thought he was prepared for the worst. He would say later that he knew that something ugly was going to happen. Still, he added, "in my wildest imaginations, I never expected what happened."

*O*nly two members of the BYU administration attended the Utah press conference. One was Julie Walker, the electronic media expert in the public affairs department. The other was Alan Knight, an advertising sales representative for the BYU alumni magazine, *BYU Today*. Knight happened to be in Salt Lake City that morning, so he volunteered to help Walker make a tape of the announcement.

After the announcement, Knight went to pick up a copy of the press release and found himself face-to-face with Chase Peterson. Peterson said, "Gee, I'm glad somebody from BYU came up. We invited BYU to be a part of the press conference. They turned us down."

When Knight returned to BYU with this tale, it "raised some eyebrows," in the words of Paul Richards, the PR director. One assumes that Walker's tape of the conference raised a few more. At one point in the proceedings, a reporter asked whether Pons and Fleischmann were aware of any similar work going on elsewhere. "Let's see, I'll answer it,"

Brophy said. "We're not aware of any such experiments going on. There are none reported in the literature."

Down in Provo, Jae Ballif and Jeffrey Holland spoke to a gathering of the BYU Faculty of Science. The two administrators explained that agreements had been made with the University of Utah and apparently been violated and that they were going to do everything they could to protect the integrity of both Steve Jones and the institution. Said Ballif, "We wanted them to know that we were going to act honorably, and that we were going to protect the integrity of our people."

Afterward, Ballif and Holland walked to the physics building, where Jones and his colleagues were working furiously to finish the paper. By this time their various sources had confirmed that Pons and Fleischmann had already submitted a paper to a journal other than *Nature*. Since this was a flagrant violation of the agreements, they were now justified in submitting their paper to *Nature* as soon as possible. While Pons and Fleischmann were still doing post–press conference interviews in Pons's basement laboratory, Bart Czirr faxed off the BYU paper on piezonuclear fusion to the Washington office of *Nature*.

That night Jan Rafelski, who was in town to help, treated his BYU collaborators and their wives to dinner at a local Mexican restaurant. Paul Palmer would recall that it felt like a victory celebration: "We just wondered what our victory was; we decided our victory was that we were now separated from those guys."

This still left open the question of whether to abide by the now many times violated agreement to meet at the airport at two o'clock the following afternoon and send off the simultaneous submissions.

Jones said he was tired of the charade. He had planned to spend the Easter weekend in Denver with his family. He stayed home on Friday to pack.

Friday morning, Paul Richards called Palmer and said that a local television news team had heard about the airport meeting. Apparently these broadcasters pictured the joint submission as a modern-day equivalent of the golden spike ceremony, which had celebrated the joining of the east and west lines of the transcontinental railroad. The news team wanted to know if BYU was going to go through with the meeting.

Czirr and Palmer called Jones at home and asked for his opinion. Jones suggested Czirr go because he would be hard-nosed about the charade. Czirr was flattered but said he wouldn't do it. They'd already submitted the paper. Czirr observed with pungent candor that this was both "immoral and stupid." He said, "Pretend on national television, big event, old brotherly love here. They betrayed us Thursday and we've already submitted. It's a big farce. I won't go." Czirr and Palmer then called Paul Richards, who suggested maybe he would go, but Czirr badgered him out of it. That was the end of it.

The television station sent a cameraman, who filmed Marvin Hawkins waiting around the Federal Express office with an envelope in his hands.[1]

Book Two

A
COLLECTIVE
DERANGEMENT
OF
MINDS

Men, it has been well said, think in herds; it will be seen that they go mad in herds, while they only recover their senses slowly, and one by one.

CHARLES MACKAY,
*Extraordinary Popular Delusions and
the Madness of Crowds,* 1841

There is something fascinating about science. One gets such wholesale returns of conjecture out of such a trifling investment of fact.

MARK TWAIN,
Life on the Mississippi

At 4:45 in the afternoon on March 23, Al Bard found a note from his secretary that Mark Blackwell of the *Dallas Times Herald* had called about "Pons and Fleischmann fusion."

Bard was a professor of chemistry at the University of Texas in Austin and was considered by many to be the best electrochemist in the country. He was the editor of the *Journal of the American Chemical Society,* which was considered the premier scientific journal in the field. Bard was a natural source for reporters who were looking for an informed opinion on a chemistry story, and he knew both Stan Pons and Martin Fleischmann well. He went back thirty years with Fleischmann, and when Utah had hired Pons, the chemistry department had asked for a recommendation from Al Bard, who wrote that Pons was a "very good grab" for a chemistry department like Utah's, which was one of the top twenty in the country.

But that was then. Pons and Fleischmann fusion?

"You really got this screwed up," Bard said to his secretary after reading the message. Then he phoned Blackwell, who described what he knew of the announcement out of Utah.

"Did they see neutrons?" Bard asked. "Did they detect tritium?"

Blackwell said he would find out. An hour later he called back. Yes, Pons and Fleischmann reported neutrons, and tritium; they detected more heat coming out of the cells than could be accounted for by the energy going in.

"So," Blackwell asked, "what do you know about Pons and Fleischmann? Do you believe this?"

"They're both respectable scientists," Bard said. "It's an amazing effect. If it's true, it's terrifically important."

This is what scientists invariably tell reporters when referring to seemingly bizarre discoveries of which they have no knowledge. Indeed, Bard didn't know what else to say. It seemed so strange.

2 COLLEGE STATION, TEXAS, AND PASADENA, CALIFORNIA

Chuck Martin was driving home from his laboratory through the miles of student housing that seem to constitute the better part of College Station, Texas. He was listening to National Public Radio's *All Things Considered* when the newscaster reported that two Utah chemists claimed the discovery of nuclear fusion in a test tube. Martin all but ran his car off the road.

Martin was a rising electrochemist at Texas A&M and a close friend of Stan Pons. He considered Pons not only a very good scientist, maybe even a genius, but a generous and thoughtful man. Just two days earlier, Martin had spoken to Pons, who had told him to watch the news on Thursday, the twenty-third. Martin had forgotten. Now he remembered: "I can't believe this. This is what Pons was talking about. My God!"

Martin tried to call Pons but couldn't get through. He then tried Nate Lewis in Pasadena. Lewis, at age thirty-three, was two years younger than Martin, but they had come up through the ranks of academia together and had known each other since 1982. Lewis was an electrochemist at the California Institute of Technology and was considered by his colleagues to be a prodigy of a sort. Martin put it this way: "I'm not as young as Nate, and nobody's as smart as Nate."

Unfortunately, Lewis was on a flight back from Washington, D.C. His wife, Carol, told Martin that he wouldn't be back until late—perhaps ten o'clock. Martin told her to tape the Cable News Network, which was broadcasting excerpts of the Utah press conference. He called back at nine o'clock; Lewis was still en route. Finally Lewis returned, and his wife relayed the message, along with the editorial opinion that Chuck Martin had gone crazy.

Lewis watched the videotape and then called Martin, who was already compiling a list of the equipment required to induce cold nuclear fusion. To Martin, cold fusion represented the opportunity of a lifetime, manna from the heavens. "How many times in your life," he said, "do you have the chance to do an experiment that could impact all of society? This is potentially the most important experiment of the twentieth century. And I've been an electrochemist for a decade, and all of a sudden electrochemistry is the most important issue in science. And I can make a contribution. My God, it has fallen right into our laps."

"I'm going to go do this," Martin said. "I can't wait. It's unbelievable."

"Well," said Lewis, who was not so tight with Pons, and thus wanted to see a published scientific paper before he bought into it, "I'm not going to do it until tomorrow."

3 CAMBRIDGE, MASSACHUSETTS

Half a dozen young chemists were drinking at the campus bar at the Massachusetts Institute of Technology when Richard Crooks, who was a postdoctoral fellow, pointed to the television screen, which was, as usual, turned to the news. "What's Stanley Pons doing up there?" Crooks asked.

These chemists worked for Mark Wrighton, chairman of the MIT chemistry department, and all of them had heard of Stan Pons. They had also finished three or four pitchers of beer already that evening. What followed was a story they would tell many times over the next few years.

The pub was loud, and the volume was off on the television, so they started shouting to the bartenders, "Turn up the volume! Turn up the TV!" But the bartenders couldn't hear them. So they sat and watched as the images of Pons and Fleischmann flashed on the screen, followed by pictorial diagrams of fusion, two billiard balls coming together, fusing. And they all looked at one another. Pons and Fleischmann? Fusion? What was going on here?

One of them said, "Well, if it's important we'll find out about it tomorrow." Then they returned to their beers.

4 THE PRESS

Friday morning, *The Wall Street Journal* page one headline was nearly impossible to miss: TAMING H-BOMBS? TWO SCIENTISTS CLAIM BREAKTHROUGH IN QUEST FOR FUSION ENERGY. IF VERIFIED, UTAH EXPERIMENT

PROMISES TO POINT THE WAY TO A VAST SOURCE OF POWER. Then below this: BATTERIES AND PALLADIUM WIRE.

> SALT LAKE CITY—Scientists working at the University of Utah made an unprecedented claim to have achieved a sustained hydrogen fusion reaction, thereby harnessing in the laboratory the fusion power of the hydrogen bomb.
>
> The two scientists said that with no more equipment than might be used in a freshman chemistry class, they had triggered a fusion reaction in a test tube that continued for more than 100 hours.

The story mentioned only briefly that the University of Utah had not released a scientific paper with their announcement, which meant that the scientific community had no way to make an informed assessment of the results. Of those who had ventured uninformed opinions, the article only said that many "expressed a gut reaction of incredulity."

From day one, *The Wall Street Journal* would be first and foremost in cold fusion. It was rumored that the *Journal* editors felt they had been embarrassed by *The New York Times* on the room-temperature superconductor story, which had broken two years earlier. Thus, the editors elected to win this one at all costs. Within a month, the *Journal*'s seeming omnipotence would have scientists calling it, not without some displeasure, *The Wall Street Journal of Physics* or *The Wall Street Journal of Cold Fusion*.[1]

In fact, the *Wall Street Journal* staff had been so smitten with the cold fusion story that they were surprised Friday morning to find that their competition at *The New York Times* had run the story not on page 1 but at the bottom of page 12 under the headline NUCLEAR POWER GAIN REPORTED BUT EXPERTS EXPRESS DOUBT. When the *Journal* editors saw the *Times*'s treatment, they promptly phoned their bureau in Los Angeles and asked how the *Los Angeles Times* had played it. They were told the *L.A. Times* had also run cold fusion on the front page (PAIR PROCLAIM NUCLEAR FUSION BREAKTHROUGH . . . SCIENTISTS IN UTAH SAY SIMPLE TABLE-TOP DEVICE PRODUCES MORE ENERGY THAN IT USES IN TESTS).[2] "Thank God," Jerry Bishop said. "We thought we had our necks stuck way out."

Bishop saw the story as his own. He might have played it down, as the uptown competition had, if not for the unorthodox method of announcement. After all, this was an affair endorsed and promoted by the president of the University of Utah. "It was kind of secondary whether it was true or not," Bishop said. "Just the fact that the claim was made under the conditions that they made it. If Pons had called me directly, I would've said, Go away."

Meanwhile, *The Washington Post* ran a cautionary story on page 3: SIMPLE TEST SUSTAINED FUSION, SCIENTISTS SAY. PHYSICISTS DOUBTFUL OF PRACTICAL APPLICATIONS. Philip Hilts, the *Post* reporter, communicated his personal skepticism in the third paragraph: "Other scientists expressed doubt about the work, however, and one physicist familiar with it said the announcement was 'blown far out of proportion.' "

What was recurrent in all these first cold fusion stories was the obvious comparison between Pons and Fleischmann's absurdly simple contraption and the mammoth multimillion-dollar efforts to induce fusion with conventional methods.

The *Journal* quoted an unnamed source at the University of Rochester saying that the conventional laser fusion program there might "ignite a hydrogen pellet within three to four years." Cost to the taxpayer of this "inertial confinement" approach to nuclear fusion: $150–160 million a year.

The *Los Angeles Times* noted that the Department of Energy was spending $350 million in research on magnetic confinement fusion, which was the other conventional approach to fusion power. With all this cash, scientists at Princeton's Plasma Physics Lab "were able to generate temperature ten times hotter than the core of the sun, about 360 million degrees Fahrenheit, but they were able to hold it there for only a fraction of a second." Charles Barnes, an astrophysicist from the California Institute of Technology, noted in the *L.A. Times* that "a few thousand dollars will get you a few quarts of heavy water and a few hundred will get you the palladium and platinum." He seemed to be saying, What are we waiting for?

After the Utah press conference, the reporters had been invited to Pons's basement lab to view the fusion reactors. These were situated in a back room of the four-room laboratory, sandwiched unheroically between a janitor's closet and a utility room. The apparatus was less than awe-inspiring, but, as the press release had said, about what one would expect from a freshman chemistry course. A stainless steel water bath, the size of a large picnic cooler, sat on a table in the center of the room. Inside that were two fusion cells, sophisticated test tubes, submerged neck-deep in the water, wires emerging to connect the electrodes inside to a stack of power supplies. Another three fusion cells sat off to one side, bubbling away like freshly poured soda water. That was it. All five fusion reactors and their accompanying electronics could have been loaded comfortably into the back of a station wagon. Two car batteries were even sitting in plain view behind the apparatus, giving the distinct impression that the power supplies for room-temperature fusion could be purchased at Sears or Pep Boys.

5 SALT LAKE CITY, PROVO, AND POINTS SOUTH, UTAH

At the public relations office at the U, a kind of euphoric madness reigned. The office received over 400 phone calls in the first days of cold fusion (or over 400 phone calls an hour, depending on the source) from, among others, major corporations, investors, management consultants, booking agents, well-wishers—"Congratulations on your incredible discovery. We are very happy to be alive during this time in humanity's history, and we are looking forward to experiencing a world free of pollution, etc. etc."—old friends, new friends, senators, congressmen, foreign consuls, publishers hoping to sign Pons for his memoirs, filmmakers, not to mention Texans fearing the discovery would destroy their already shaken oil-based economy, and, of course, scientists and reporters.

Friday morning, Chase Peterson attended a scheduled meeting of the University of Utah's board of regents. The timing was serendipitous to say the least. Peterson gave a brief report on cold fusion, at which point Ian Cumming asked, "What's next?"

Ian Cumming may have been the state's single most influential citizen. He was a community benefactor who shunned the press, sat on boards, and gave pep talks to luncheon clubs on the Utah economy. His personal point of view was "Utah must do more—spend more if necessary—to encourage job development in the state by attracting industry from without and cultivating industry within." It was said around town that Cumming would be the man to choose the next governor of Utah.

"What do you mean what's next?" Peterson said.

"What's the next step?" Cumming asked.

"Challenge and confirmation," said Peterson, "and we'll know in six months or a year what goes on here."

"We can't wait that long," Cumming insisted. "We can't let this thing get out of our hands. We've got to keep centrality to it in Utah. And so we need to get some money from the state to get this thing moving promptly."

They immediately phoned the governor, Norman Bangerter, who was relaxing at his vacation home in southern Utah. Cumming then had his private jet pick up the regents in Provo, and they all flew south to meet with Bangerter. It was an impulsive gesture, which, in the words of Hugo Rossi, "would torque the whole affair up several significant notches."

The regents, led by Cumming and Peterson, unfolded this great discovery for the governor. After they discussed it for twenty minutes, he said, "Okay, you didn't come all this way just to tell me about it. How much money do you want?" Bangerter's guests looked momentarily startled. Then one of them suggested $2 or $3 million might do it. Bangerter offered $5 million, "if this is as good as you say."

This seemed to be the governor's attitude about cold fusion throughout: "Knowing nothing about it," he would say "I am highly optimistic."[3]

Under the circumstances, Bangerter's $5 million offer may be considered cautious. Miles Crenshaw, for instance, a local personality and radio talk-show host, suggested the state divert $300 million to cold fusion research, with $1 million tax-free to Pons and Fleischmann as an expression of gratitude. Another state senator quickly suggested that citizens be able to invest a portion of their taxes in cold fusion research, to be paid back at a later date through the proceeds reaped by the state. It was a heady time.

6 AUSTIN

"Friday," Al Bard said, "all hell broke loose."

Bard had lived in the heart of Texas for three decades, but his native Bronx accent hung on stubbornly. He had gone to Bronx High School of Science, City College of New York, and then Harvard, where he first studied inorganic chemistry, which was then a "hot" field. He studied the discipline under a young British assistant professor named Geoffrey Wilkinson, and even a graduate student could tell he was doing hot work. Harvard, however, announced that Wilkinson would not be given tenure, and Wilkinson left shortly thereafter. (In 1973, Wilkinson shared the Nobel Prize, proving that Harvard could be as fallible as any other university.) Bard, meanwhile, had to find a new mentor. He got "a little bit interested" in electrochemistry, and James J. Lingane signed him up.

Lingane had been the student of I. M. "Pete" Kolthoff, who was the father of electrochemistry in America, at the University of Minnesota. Under Lingane, Bard learned to use electrochemical techniques as tools to attack other interesting chemical problems. This made Bard's reputation, and, with his peers, it launched a renaissance in electrochemistry, which happened to coincide with the phasing out of those departments at schools like Harvard and Princeton. This explained, among other things, why Bard went on to Texas.

Bard spent much of Friday, March 24, trying to glean what facts the news reports provided on cold fusion, which he then recorded in his lab book. He learned that Pons and Fleischmann had worked on cold fusion for five years, that it used equipment along the lines of that in a freshman chemistry experiment, that the sophisticated test tube was the size of a drinking glass; that it used 99.5 percent heavy water, that the reaction took ten hours to get going and continued for more than a hundred hours, past the point at which the power consumed was greater than the power generated—what physicists would refer to as break even—and that it produced four watts of power. *The Wall Street Journal* and the *Dallas Times Herald* both confirmed that Pons and Fleischmann had detected neutrons and tritium. According to the *Austin American-Statesman,* any electrochemist could reproduce the experiment in an afternoon. It looked, Bard thought, like a snap.

Bard then wrote down questions to which he wanted answers: for instance, how could there be sufficient pressure within the palladium to cause fusion? A deuterium nucleus is composed of a single proton and a single neutron, so the nucleus has an overall positive charge. That means two deuterium nuclei would naturally repel each other. In order to induce fusion, this repulsive force has to be circumvented or overwhelmed, which can be done with sufficient pressure and heat, as in the cores of stars. In the 15-million-degree plasma at the core of the sun, for instance, the nuclei are moving so fast that they smash into one another and fuse without hesitation. But how could this be achieved chemically? How could a palladium rod and a small electric current force the two deuterium nuclei to come together? It didn't make sense.

At noon Bard met with his research group to tell them what he knew, and that he was going to do cold fusion experiments: "It's not that we're going to make any big contribution now," he said. "It's done. If it's true, it's true, but it's important to reproduce this result." Indeed, Bard felt as he had when he was young and had read about a classic scientific experiment and wanted to reproduce it. Not because he wanted to write a paper about it but because he wanted to do that experiment with his own two hands. "I think what most of you are doing is important enough," Bard said to his researchers. "I don't think you guys ought to jump off and get into this unless you really feel very strongly motivated about it. And besides that, this experiment is potentially very dangerous, as far as I can figure out. So the lab where I'm doing this is going to be off limits." And Bard added, "I don't want you to feel that I'm doing this all for myself. If you guys want to do it, you're welcome to."

The graduate and postdoctoral students seemed to think that cold fusion, even with all the hullabaloo, was improbable at best. They didn't want to get involved. One postdoc, Norman Schmidt, had just joined the group, and he volunteered. So it was Bard and Schmidt alone.

That Bard, in his late fifties, actually ran the experiments himself was unusual. Most chemists of his age and caliber consider themselves managers of a sort and believe that the laboratory is a place for graduate students and postdocs. So did Bard, until cold fusion.

Bard tried to call Stan Pons, but he couldn't get through.

At 2:00 P.M., Bard received a call from Larry Faulkner, who was dean of the College of Arts and Sciences at the University of Illinois and had studied electrochemistry under Bard. Faulkner said that he didn't believe the cold fusion announcement. At 5:00 P.M. Bard heard from Mark Wrighton of MIT, who said he was withholding judgment, but the MIT physicists, on the whole, were skeptical. Wrighton said that they were worried about neutron radiation. If Pons and Fleischmann were producing four watts of fusion power, then where was the radiation that not only is a sign of fusion but would have constituted a serious health hazard?

Nuclear reactions generate nuclear radiation. That is the nature of the beast. Even without calculating exactly how much neutron radiation should have been emitted by four watts of fusion power, one indication that Pons and Fleischmann had observed too little was their reasonably robust appearance.[4] The radiation emitted from this level of power generation should have been sufficiently malign for its effects to have been noticeable. In fact, the radiation would have killed the two chemists, not to mention seriously impaired the health of the students working nearby. But it hadn't. So where were the neutrons?

Bard and Schmidt ran a quick and dirty experiment that night. They placed a palladium rod and a platinum rod in a flask of heavy water, hooked the two rods to an electric current, and compared the results of that setup with those of a similar flask in which both electrodes were made of platinum. The idea was that if the deuterium fused in the palladium electrode, that cell would produce more heat than the cell that had two platinum electrodes. What Bard didn't know was what manner of electrolyte Pons and Fleischmann had used. He tried D_2SO_4, a heavy water variation of sulfuric acid. He and Schmidt ran both cells off the same power supply with about the same cell voltage, then looked for excessive heating in the palladium cell. "It would have to be something very big to see it," Bard said. "We didn't see anything."

Bard was marginally skeptical. He agreed with a philosophy that Fleischmann would later articulate: "that if you really don't believe something deeply enough before you do an experiment, you will never get it to work." It may sound like the kind of instructions the Good Witch gave Dorothy to get her home from Oz; nevertheless, one has to have some faith to get things going.

By Saturday, Bard was calculating possible electrochemical routes to achieving fusion, searching for some understanding of what conditions might lead to very high pressure in the palladium. So he estimated how high the pressure would have to be to induce fusion, and came to 10^{25} atmospheres, which is only a factor of 100 less than the 10^{27} atmospheres Fleischmann claimed they had achieved at Utah. This may have been an interesting coincidence, but it still seemed absurd. 10^{27} atmospheres is one of those enormous numbers that scientists bandy about with such a cavalier spirit: one billion billion billion times the pressure of the atmosphere on the surface of the earth. To put this enormous number in perspective, it is roughly 10 million billion times the pressure at the center of the sun. How could it be possible to produce even a few million atmospheres in any earthly apparatus without that apparatus being crushed or exploding?

So Bard and Schmidt considered the possibility that microelectrodes— or rather ultramicroelectrodes—had to be used. Fleischmann and Pons had lately been working with ultramicroelectrodes, so this was not idle speculation. Bard thought that maybe with these "tiny, tiny electrodes," they could induce a stupendous current density across an electrode and, in one instant, create an enormous pressure in the palladium. So they took twenty-five-micron wire and built ultramicroelectrodes of palladium. But they failed to detect any evidence of fusion.

When Nate Lewis arrived at his Caltech lab on March 24 at a little after 8:00 A.M., two of the postdocs—Mike Sailor and Reggie Penner—already had a cold fusion experiment on the verge of running. Lewis was skeptical. He told Penner and Sailor that the experiment wasn't worth more than one day's effort. It was also possible that Lewis, having a high opinion of his own talents, did not want to sweep up after Stan Pons and Martin Fleischmann, even if they had just discovered the greatest thing since fire.

Lewis was proud of this reputation. "You can go out and get a reading on me. I did things and got the respect of a fair number of people by going into an area that people have studied already, and doing things meticulously and seeing things that other people didn't see. So I'm one of the people, where you say, 'If Nate says it's right, it's right.' " Lewis had done his doctorate at MIT under Mark Wrighton, who had been electrochemistry's resident prodigy before Lewis came along. Then Lewis went to Stanford as an assistant professor without bothering with a postdoc. Now he was thirty-three, and tenured at Caltech.

Lewis preferred a blend of scientific talents in his group. He believed that if he put an inorganic chemist in with an organic chemist in with a physical chemist and so on ideas from the different disciplines would cross-pollinate. His style was to publish eight papers at once on a single subject, saturating it with these various disciplines. Thus, what Lewis published tended to be the last word. Chuck Martin said of Lewis: "The thing you've got to realize about Nate is, underneath his calm exterior, he is one of the most fiercely competitive son-of-a-bitches you'll ever meet. He's very focused, and he relentlessly pursues things."

Because of Lewis's reputation and his science, he attracted postdoctoral and graduate students who wanted to pursue the most challenging science and thought they were good enough to handle it. In 1989 many of these happened to be Stanford émigrés who had come down with Lewis a year earlier, when Lewis jumped schools. Lewis gave them freedom to do what they wanted. Their long-term projects then evolved into publications. Lewis acted as mentor, impresario, and *arbiter scientiarum*. In the words of Mike Heben, one of the Stanford émigrés, Lewis would "filter through the bullshit and find out what's right and what's not."

Mike Sailor had arrived in town with his Ph.D. from Northwestern before gravitating to Lewis's group. He was a synthetic chemist who had worked on metal cluster compounds. He said Lewis's lab looked "like a fun place to spend a couple of years."

Reggie Penner had done his Ph.D. with Chuck Martin at Texas A&M. He was currently working with Mike Heben on a scanning tunneling microscope project. A commercial STM would have cost about $70,000, so Heben and Penner had built their own. "You can't buy a commercial instrument that will meet our requirements," Penner said. With this hand-built STM, they had achieved atomic resolution, which is to say they could make out single atoms under the microscope. It was like a novice race car driver building his own Formula I because a Ferrari wouldn't meet his needs.

Penner and Sailor had both studied the lengthy coverage of cold fusion in the *Los Angeles Times*. Not being nuclear physicists, they didn't know much about fusion, so they lacked a certain skepticism. In fact, they were, as Penner would say, blown away by the news. He was even a little angry that he hadn't thought of the idea himself. They were familiar with both Fleischmann and Pons, whom they knew to be pioneers in micro-electrode research. Penner estimated that he had forty some papers in his files written by one or the other or the two together. He had met them both.

When Lewis walked into his lab on the morning of the twenty-fourth, Sailor and Penner were assembling their first cold fusion cell. They had bootlegged a strip of palladium and purchased the heavy water from the stockroom—sixty dollars for one hundred grams. They had assumed Lewis could afford it.

Curiously, neither Sailor nor Penner expected to succeed. They assumed that the experiment had to be considerably more complicated than the *Los Angeles Times* had made it out to be. Sailor, however, wanted to tell his grandchildren that he had tried cold fusion the day after the discovery was announced.

As the day went on, the Caltech experiment went through various stages. They ran a control cell by replacing heavy water with ordinary water, the idea being that if excess heat was produced by fusion of the deuterium molecules, then ordinary water, which is only $1/3000$ deuterium, would certainly induce less fusion and less heat.

This is the most fundamental commandment in the canon of experimental technique. To reach an unimpeachable conclusion establishing the cause of an effect, run controls. E. Bright Wilson, Jr., in his classic 1952 volume *An Introduction to Scientific Research,* described controls as "similar test specimens which are subjected to as nearly as possible the same treatment as the objects of the experiment, except for the change in the variable under study."[5]

The cell with ordinary water seemed to behave identically to the cell with heavy water. Neither of them seemed to have induced a noticeable nuclear fusion reaction.

Because they expected neutrons and gamma rays to radiate vigorously from the cells if they induced fusion, they hung up radioactive warning signs and borrowed Geiger counters from a radiochemistry lab. They also placed the incipient fusion cells on top of Polaroid film, under the mistaken assumption that if gamma rays were emitted they would expose the film. Sailor said later, "Gamma rays would go right through the stuff, but we didn't know that." Nothing happened.

They joked about lining their shorts with lead foil. "Neutrons would go right through that stuff, too," Sailor said. "But we didn't know that either."

Meanwhile, Lewis talked to John Gladysz, a chemist from the University of Utah who was on sabbatical for the semester at Caltech. Gladysz had known Pons for seven years and told Lewis that he'd "bet on Stan." Lewis also decided that Pons didn't look like someone who "went off the deep end, . . . didn't look like he was really incoherent." Still, they might have quit right then had not a synergistic effect begun with the physics department.

Early in the day, Steve Koonin called Lewis. At age thirty-seven, Koonin had as formidable a reputation in nuclear physics as Lewis had in electrochemistry. Koonin was on sabbatical from Caltech at the University of California at Santa Barbara and called to find out what Lewis knew of cold fusion and these two guys Pons and Fleischmann.

"They're a little flaky," Lewis said, "but they're not crazy. I don't know what to think."

In return, Koonin told him that there already seemed to be an independent confirmation. He told him that Steve Jones at BYU had apparently detected neutrons emitted from the kind of electrolysis cells that Pons and Fleischmann had used. Koonin had chaired the JASON panel on Jones's muon-catalyzed fusion work, although Jones had never mentioned cold fusion at the time. But Jones had been mentioned in the *Los Angeles Times* article that morning:

Physicist Steven Jones of Brigham Young University in nearby Provo is widely recognized for his work in a similar field which involves the use of subatomic particles called muons to create a fusion reaction. That generally is regarded as a promising area of research, but Jones indicated in a tele-

phone interview Thursday that he has switched his research to the same area
now being studied by Pons and Fleischmann.

Jones, who was reluctant to even discuss his work until a formal paper is
published in May, said that after several years of frustration, "we're seeing
something significant."

It certainly appeared as if Jones, a recognized name in fusion physics, had
seen Pons and Fleischmann's work, found it compelling, and replicated
it. The *Times* also quoted Jones saying, "It's a scientific success," which
seemed definitive.

Koonin suggested that Lewis connect with Charles Barnes and the
physicists at the Kellogg Laboratory, the nuclear physics lab at Caltech.
Lewis suddenly began looking at the cold fusion scenario as not quite so
implausible.

So he called Barnes, a sixty-seven-year-old astrophysicist whose forte
was studying exotic fusion reactions in stars and who was on the verge
of calling the head of the chemistry department to get a line on a good
electrochemist. Barnes had already discussed cold fusion with two of his
research fellows, Steve Kellogg and T. R. Wang. It seemed that Barnes
and his colleagues had one of the best neutron detectors in the world, and
if they could learn the proper electrochemistry, they could set up a cell
and be in business.

So Lewis made the arrangements with Barnes, then told Penner and
Sailor that they had a neutron detector at their disposal. The two postdocs put their cells on a pushcart and wheeled them across the campus.
Kellogg and Wang then inserted the cells into the center of the neutron
detector, which they called the neutron polycube. It was made up of a
dozen neutron detectors arranged in a twelve-inch-diameter cylindrical
array. They all looked inward on a four-inch-square borehole. The
entire contraption, complete with paraffin shielding to slow down any
neutrons, was about the size of a phone booth or a shower stall.

Kellogg and Wang placed the cells in the borehole and rolled the
detector into its chamber. Then they waited. The cube was 100,000
times more sensitive than the neutron detector Pons and Fleischmann
had used. So if these Caltech cells emitted any neutrons at all, the cube
would have little trouble detecting them. It observed nothing.

Meanwhile, Lewis spent much of the day calling around the electrochemistry community trying to learn the exact ingredients in Pons and
Fleischmann's electrochemical cells. He was primarily concerned with
the nature of the electrolyte. Penner and Sailor were using perchlorate,
and if that wasn't part of the Utah recipe, he wanted it out. "It's like

rocket fuel," Sailor observed. "There are documented cases of perchlorate exploding on people, people losing eyes and fingers. There's a lot of chemical energy in perchlorate."

Lewis had been trying to reach Pons all day; by Friday evening he managed to get hold of Jim Brophy. Lewis told Brophy that they were trying to replicate the work; they needed help, and they would love to confirm this. He said he knew Stan Pons personally, and he added that they had one hell of a sensitive neutron detector, and they were already going full blast. Brophy said he'd see what he could do.

Lewis even sent Pons a message through the electronic mail network, or e-mail for short:

Date: Fri, 24 Mar 89 18:25:16
From: pgs%Xray. Caltech.
Subject: FUSION
To: pons

Hi stan.
I am sure that you have been inundated with requests, but we have an extremely sensitive neutron counter and can confirm the reaction, however our initial attempts . . . did not yield any neutrons. Could you please either send me a preprint and/or bitnet explicit directions of electrode preparation, D electrolytes, voltages, etc., so that we can confirm the experiments and quantitate the neutron yield? The particle physicists here have been very helpful and we can certainly lend credence to the proposal if you would provide us with explicit directions. Thanks for your help. . . .

Pons called Lewis back that night around 8:30. The Utah chemist sounded relaxed on the phone, surprisingly affable. The conversation, however, was a peculiar one.

"Hey, I'd like for you to confirm this," Pons said.

"We'd love to do it," said Lewis.

"I'll send you a preprint," Pons said. "But I can't tell you about it now."

"Why not?"

"I can't," Pons said. Then he added, "We don't think there's many neutrons. This is a nonclassical nuclear reaction."

"Oh."

"Don't worry," Pons said. "You'll figure it out. Look for the heat. The heat's the ticket. And I'll send you a preprint when we can. . . . Be careful, we've had explosions. I don't want you to hurt yourself. I want you to be careful."

"Well, thanks a lot," Lewis said, trying to sound sincere. "Well, what can we do wrong? What should we avoid? Because I don't want to hurt myself."

"Avoid high currents and sharp edges," said Pons.

"Okay," Lewis said. That was it.

By the end of the day, Lewis's chemists and Barnes's physicists agreed that they had barely enough information to make it worthwhile continuing. If any neutron radiation had been emitted by their cold fusion cells, they'd have seen it. However, if they had been on the verge of quitting earlier, by Friday night it appeared that cold fusion had its hooks in these Caltech scientists. Now, heaven help them, it seemed to be a matter of pride.

Saturday morning Sailor started building an apparatus called a calorimeter. Pons had said look for the excess heat. The calorimeter would measure the heat output from the cells.

8 CAMBRIDGE

The first cold fusion attempt at the Massachusetts Institute of Technology came from a disenfranchised fraternity turned cooperative living group. On Friday morning they circulated copies of an announcement that included a photograph of a glass jar that looked as if it might once have held strawberry preserves. Now all one could see was a metal rod, or maybe two, emerging from the top. This jar sat in a tub of water, which sat within a pile of bricks. Beneath the photo, it said:

> Here's a photo of MIT's first *attempt* at a room temp fusion reactor, manufactured in 15 minutes last night. We did heavy water (99.8%) electrolysis with two palladium electrodes, temp in vial went from 20 degrees centigrade to 50 degrees centigrade. Doesn't mean shit since the electrolysis process itself is exothermic. Sorry about the overexposed photo taken with an oscilloscope camera. That's all we had available. (The lead bricks were for neutron shielding.) Just some undergraduates having fun. . . .

It was signed by a member of Pi Kappa Alpha, or pika, as the coop now referred to itself.

Marcel Gaudreau, a nuclear engineer at the MIT Plasma Fusion Center, was handed a copy of the pika announcement on Friday morning. When he read it, he started to take this cold fusion business seriously.

Gaudreau stopped in to see his boss, Ron Parker, who was the director of the Plasma Fusion Center. Parker had already heard about cold fusion. He thought Pons and Fleischmann were what he called "squirrels from his nut file."

The Plasma Fusion Center, known as the PFC, is one of the half-dozen research centers in America dedicated to research in conventional thermonuclear fusion. It is located behind the MIT campus, in a part of Cambridge made up of reconstituted candy factories. Parker oversaw a $25-million-a-year program aimed at achieving a sustained fusion reaction in a tokamak, an immense donut-shaped fusion reactor.[6] He expected that MIT's next generation of tokamak would reach break even, which is the point at which the device would generate more energy than it took to keep it running.

Parker was a serious man, despite a distinct resemblance to the late British comedian Peter Sellers. He had invested his life in fusion power and believed that the only things needed to turn it into a viable energy source were hard work and dedication. He hoped he would see success in his lifetime. Gaudreau was only slightly less serious, a frenetic French Canadian who had also devoted his life to fusion.

At the time he met with Parker, Gaudreau's working hypothesis was that Pons and Fleischmann's little electrolytic cell was just a display model they could show the press. Gaudreau considered this a variation on what he called the Christmas tree effect: "Say for example you want to bring the president to the control room, turn all the terminals on. That's called the Christmas tree effect, and the more light the better." So this cell existed only to demonstrate the simplicity of the device. Meanwhile, somewhere in their basement laboratory, Pons and Fleischmann had sequestered a huge reactor surrounded by tons of concrete shielding. If this were true, it would explain how cold fusion might have generated the requisite neutron radiation without having proven detrimental to the health of the two chemists.

Gaudreau told Parker, "Ron, I've got enough things in my life right now. I don't need anything more. But if you need someone to look into cold fusion, I'm ready."

Gaudreau spent Friday night viewing videotapes of the news reports. He also called the pikas. He offered them "legitimacy" for their experiments. He'd give them a laboratory, neutron detectors, and safety equipment. In return, he insisted that they refrain from running any more cold fusion experiments until he could check them out. Gaudreau wanted assurance that the pikas weren't frying themselves and maybe even their neighbors with neutron radiation. The pikas agreed.

First thing Saturday morning, Gaudreau stopped by the former fraternity house. He picked up the cold fusion jar and put it in a lead-lined bucket to carry it back to the PFC. We don't know what this thing is doing, Gaudreau thought.

While Gaudreau gingerly walked across Cambridge with something that for all he knew might be a small untactical nuclear weapon, Mark Wrighton was meeting with a handful of his electrochemists and grudgingly deciding that they should pursue cold fusion.

Wrighton was another legendary figure in electrochemistry. His career had been almost preternaturally accelerated. He was the son of a navy man and had moved around a lot as a youth. He was born in Jacksonville, Florida, and was schooled in Virginia, Tennessee, Maryland, and Newfoundland. He had studied chemistry at Florida State, graduated in December 1969, took up at Caltech in January 1970, and two years later, at age twenty-three, he had his doctorate and was already an assistant professor at MIT. By age twenty-eight he was the youngest tenured professor in the history of MIT, and ten years later he was head of the chemistry department. In the meantime, he had pioneered various approaches for converting sunlight to electricity or chemical energy. His peers said it certainly wouldn't surprise them if Wrighton came home with a Nobel Prize someday. In public, Wrighton appeared rational, deliberate, and sometimes cold. And, like Nate Lewis, who had studied under Wrighton, when Wrighton published, his papers were expected to be definitive.

Wrighton's researchers were simply confused by the Utah announcement. They knew that the palladium-hydrogen system, as it was called, had been studied for years by electrochemists, who had worked on it for methods of storing hydrogen, for cars that run on hydrogen. Physicists had studied it for separating tritium and deuterium from nuclear reactors. The system was so well known that it was hard to imagine any new discovery coming out of it, let alone one of such magnitude.

While they were discussing it, Martin Schloh, a Belgian graduate student, walked in and said he'd already done the experiment, which seemed to make up their minds to look into it. Schloh was on the verge of finishing his doctoral work on microelectrodes, which meant he was familiar with the work of Pons and Fleischmann. His father had flown in from Europe on Friday and over breakfast shown Schloh a copy of the

Financial Times article, which had more of the experimental details than any of the American publications.

Schloh then went back to his lab and began setting up the experiment. Friday night he wrote in his logbook, "I'm starting the experiment now," in case the experiment blew up and took him with it, although he didn't really expect that to happen. His setup was crude. He was looking to induce a meltdown. He put power to his cell, stood back, and nothing happened. He tried it again with a Geiger counter hoping to detect neutrons. He spent much of Friday night and into the morning sitting by his cell, waiting patiently for the Geiger counter to announce that neutrons had been emitted.

With this news, Wrighton called Parker, who said that they were about to give cold fusion a try at the PFC and would greatly appreciate the assistance of experienced electrochemists. So Schloh, Vince Cammarata, and Dave Albagli, both graduate students, and Dick Crooks, a postdoc, loaded up a car with power supplies, electrodes, recorders, various bottles of electrolytes, and odds and ends of electrochemistry supplies and drove the half mile to the PFC.

"We wanted to be the ones to prove it right," said Cammarata later. "We went in wanting to believe it, but being skeptical enough to say that if we see something we're not going to believe it immediately. Although the physicists were saying there weren't enough neutrons, well, that was a convincing argument. But we knew that Pons and Fleischmann are not physicists, so we knew that any measurement that they made, especially a nuclear measurement, is not going to be the best measurement possible. So maybe they could miss some neutrons. They said this could be some hitherto unknown nuclear process. Who knows? If it is an unknown process, maybe it doesn't produce neutrons. You can always rationalize anything, and you could rationalize how there would be nuclear fusion, and why [Pons and Fleischmann] wouldn't die. There are no precedents for it. One way or the other there has to be a definitive proof, and we wanted to be the ones to definitively prove it."

By Saturday afternoon as many as twenty physicists and chemists were working away on a single fusion cell at the PFC. Another dozen people had drifted by to watch. Gaudreau set up a neutron detector and a few Geiger counters to monitor radiation. They asked everyone to move away from the cell and hit the power. Nothing happened.

By evening MIT's radiation safety officers had dropped by and roped off a safety zone around the cell. They gave instructions: nobody goes within the rope while the cells are running. Gaudreau and company now set up the cell within an altar of lead bricks, beneath which they put the neutron counter. They had a meter connected to the counter, which would move if neutrons were detected. Then an undergraduate with binoculars read the meter from behind the safety rope. His instructions were to pull the plug immediately if he observed the cell going critical.

The cell didn't go critical, but that didn't overly discourage them. After all, why should they succeed? They had no idea what the current should be, the voltage, the electrolyte, anything. Both Wrighton and Parker had tried to call Stan Pons to get answers, but they hadn't gotten through.

Parker and Gaudreau speculated that they needed to create a sheet of plasma between the electrolyte and the palladium. "We thought," Gaudreau said, "what if we were to electrolyze until the palladium was fully saturated, and then put a huge pulse on it and then drive it in? Go in there and hit it with five or ten thousand volts. So that's what we did."

By early Sunday morning, Gaudreau, Cammarata, and a few pikas were the only ones left. Gaudreau had acquired a high-voltage capacitor, and for five thousandth of a second he hit one of the cells with 7000 volts. For a brief moment, they must have felt like Dr. Frankenstein cranking the juice into his monster. It's alive! It's alive! The cell glowed with an eerie blue light, which signified that they were harmlessly ionizing the electrolyte. No neutrons emerged, however.

At six in the morning, Easter Sunday, they gave up. Gaudreau and Cammarata agreed to catch a few hours of sleep, meet at 2:00 P.M., and try again. Cammarata made it back to the PFC promptly, but Gaudreau stood him up.

At some point Saturday night Gaudreau decided that there was only one certain way to determine exactly what Pons and Fleischmann had done. "So basically on Sunday morning," he explained in his staccato style, "I went home, slept for a few hours, until eleven, got up, picked up the phone, called the airport, found out when the next plane to Utah was, got my American Express and my suitcase. Took off." Cammarata only found out on Monday when Gaudreau called him from a pay phone outside Jim Brophy's office and left him a phone number. When Cammarata asked where Gaudreau had called from, Dave Albagli, who took the call, responded "wherever he is, the area code is 801."

Utah, of course.

Although Gaudreau was thirty-six when cold fusion entered his life, his routine energy level and his overall attitude had been compared, not unjustly, with those of a hyperactive child. He had come to MIT in 1972 from a small town outside Quebec City. His father was a biology professor in a small college, and, at the time, Gaudreau spoke only French Canadian. "I can honestly say that I was just off the boat." Now he wore an MIT ring, on the inside of which, he was proud to admit, were inscribed the Greek letters Theta Delta Chi. This was the symbol of both his Boy Scout troop and his fraternity, of which he was still an active participant. Ron Parker described Gaudreau as an interesting guy to talk to, "although in a lot of ways he's not representative of MIT."

With the underclassmen whom Gaudreau referred to fondly as his fraternity brothers, he was building a full-scale tokamak fusion reactor. They had no official funding. Theta Delta Chi was, in the lingo, bootlegging a fusion reactor, an endeavor that did seem somehow representative of the highly competitive and cerebral atmosphere of MIT.

As far as cold fusion went, Gaudreau refused to have any preconceived opinion. "Forget about believing," he said. "Leave that to the church. Look at the facts. Figure it out." If Gaudreau could understand hot fusion and build tokamaks, he was damn well going to figure out cold fusion: "Because if I can't figure out these little jobs, I've got no business building machines. If I can't figure this thing out, you know, I should go into the garment business."

9 SALT LAKE CITY, GROUND ZERO

Friday, March 24, Martin Fleischmann left for England, and Stan Pons spent the weekend without his mentor in what constituted a state of electronic siege. Pons had told reporters at the press conference that he planned to go skiing, but it's conceivable that he never got off the telephone.

Many of these calls were from well-wishers, including Edward Teller, the director emeritus of the Lawrence Livermore National Laboratory. Teller was the mind behind the hydrogen bomb and the Strategic Defense Initiative, which made him a controversial figure, but he was also a great believer in technology and was always looking for the unexpected.[7] Teller thought cold fusion sounded promising, and on Friday he convened a task force at Livermore to delve into it. This news must have bolstered Pons's faith that they had done the right thing. Pons also received a call from Carlo Rubbia, who had shared the 1984 Nobel Prize in physics and was director general of CERN, the huge European high-energy physics laboratory. This was a conference call arranged by the Italian newspaper *Repubblica* between Rubbia in Geneva, Fleischmann, who was stranded momentarily in San Francisco, Pons in Salt Lake City, and *Repubblica* reporters in Rome and New York. Rubbia took the opportunity to invite Fleischmann to speak at CERN. Once again Pons saw the call as confirmation. He said that the conversation was fantastic: "The man is brilliant. He picked up immediately on some of the subtleties we were discussing."

On Monday morning, it seemed that every young electrochemist in the world wanted to work alongside the coinventor of cold fusion; every newspaper reporter wanted to interview him; every scientist wanted to know how he'd done it. The deluge forced Pons to change his home phone number almost immediately, and within days change it yet again. And he got several new fax numbers and had an unlisted number put in his laboratory so he could reach his own researchers.

With all the press and all the calls, no one ever stopped to wonder how this would affect a man who admitted to stage fright while teaching his undergraduate chemistry classes. Throughout his life, Pons had displayed a marked aversion to strangers, faceless crowds, and bureaucracies. It may have been a product of his childhood; his Waldensian ancestors had suffered through eight centuries of religious persecution, so feelings of persecution may have been endemic to their psyche. What was certain was that if Pons had a choice, he always preferred working with family and friends. He'd had his son, Joey, doing research for him, and Joey's name appeared as coauthor on several of his papers. His wife, Sheila, did much of his secretarial work. (This was to cause some embarrassment later, when it was discovered that both his son and his wife were on his chemistry department payroll, although the work they were doing seemed legitimate.) His personal lawyer, C. Gary Triggs, was one of his oldest friends. Even when it came to funding for his research—Pons relied on the Office of Naval Research for $300,000 a year—his contacts were Pete Schmidt, whom he'd known for twenty years, since they were undergraduates at Wake Forest University, and Bob Nowak, who had studied with Schmidt and had received his doctorate at Michigan under the same man, Harry Mark, who had been Pons's adviser and mentor.[8]

With cold fusion, Stan Pons became a public man, whether he liked it or not. He seemed genuinely upset about the pledge of $5 million from the state of Utah, and he kept telling friends and administrators that all he wanted was to return to his laboratory and get back to work. He told *The Deseret News* that the single word he would use to describe himself was "scared." The *News,* in turn, referred to him as "the unpresuming U. chemistry professor, whose name may someday rank along side Einstein, Edison and Newton."

Meanwhile, events at the laboratory were becoming progressively weirder. After the press conference, Pons and Fleischmann had lost their transparencies, which they assumed had been stolen. Sunday night Pons apparently realized for the first time that his cold fusion lab books had also vanished. These happened to contain no more nor less than all the priceless data on the experiments. Whether Pons believed at this point that his graduate student Marvin Hawkins had stolen them is unclear, but Pons called a young postdoc named Mark Anderson and asked if he would take over the lab work on the cold fusion research. (Pons did not tell Anderson about the lab books, only that he had decided it was time Marvin Hawkins went back to his thesis work.)

Hawkins at the time was in Park City, forty-five minutes up in the mountains, where he was skiing with his wife and children. As Hawkins tells it, he had left that morning, after Pons had graciously offered to cover his expenses. Then on Monday morning, Pons called him in Park City, telling him that he needed the lab books, and Hawkins said he couldn't get them. He said he had put them for safekeeping in his brother's safe deposit box, and he wouldn't be able to reach his brother before the end of banking hours. Hawkins was under the impression that after the disappearance of the transparencies, Pons and Fleischmann had discussed with him how to safeguard the original lab books, so he had made copies to keep in the lab and deposited the originals in the safe deposit box. After talking to Pons Monday morning, Hawkins said, he drove to Salt Lake City and left Pons a photocopy of the books and a note: "I couldn't get to the safety deposit box today. I'll call you later. Here is a complete copy of the books."

That seemed simple enough.

Then, into this escalating weirdness appeared Marcel Gaudreau, fresh off the plane from MIT, looking for a collaboration and willing to settle for anything that would give him a handle on room-temperature fusion.

Monday morning, March 27, Gaudreau walked over to the university's public relations office and from there was directed to Jim Brophy, the vice president for research, who was now the front man for the cold fusion research. Gaudreau's first question to Brophy was whether what they had seen on television was the Christmas tree effect or the real thing. And Brophy said it was real. Right there, Gaudreau was thinking, they had a "quantum leap in understanding."

At this point, Brophy and Gaudreau engaged in a guarded exchange of information, a pas de deux that seems as though it might have been choreographed by Abbott and Costello or, at least, Alphonse and Gaston. Gaudreau wanted to move on to technical questions, but neither he nor Brophy was an electrochemist, so he didn't know what to ask, and Brophy wouldn't have been able to answer him anyway. Brophy did have a copy of the paper, which he showed Gaudreau but wouldn't let him copy. Gaudreau looked at it, memorized a piece of information, and handed it back. Then he excused himself, stepped outside, and called Cammarata back at MIT.

"Listen," Gaudreau said to Cammarata, "I'm not an electrochemist but this is what I read: it says something about palladium rods, you know, the size of a pencil. Now, tell me the questions I need to ask."

"Okay," Cammarata said, "find out what's in the solution. Is there any type of electrolyte?"

Then Gaudreau hustled nonchalantly back to Brophy, who once again handed him the paper. Then back to a phone. When Gaudreau first called Cammarata, his MIT office was crowded with chemists. So Gaudreau had him move to a "secure" phone, where they would not be overheard. For some reason, Gaudreau would talk only to Cammarata, maybe because he was the only one who had gone the distance Saturday night at the Plasma Fusion Center and thus had become an honorary frat brother.

"It says 0.1 molar L-I-O-D," Gaudreau told Cammarata in another call.

"Good, good, Marcel," said Cammarata. "Okay. Now find out what current density for what voltage they use."

Eventually Gaudreau trotted over to the chemistry building looking for Stan Pons himself. He waited in the hallway outside the basement lab until Pons came out and then walked across campus with the coinventor of cold fusion. Pons was off to teach a class, which was a tribute to his effort to maintain some normality. Gaudreau said, "I remember asking Pons if I could watch his class. He said, 'I'd rather not, and you won't be impressed anyway.'"

Pons told him to come back in the afternoon, which Gaudreau did, along with Brophy and two other men Gaudreau thought "looked like lawyers. Nice pants and suits." While they waited, Hugo Rossi, dean of the College of Science, showed up. Gaudreau somehow knew that Rossi had studied at MIT.

The way Rossi remembered it, Gaudreau wanted to launch an MIT-Utah cold fusion collaboration. Rossi was anxious to get Pons to work with some established group, although not necessarily Gaudreau's. Rossi said Pons was put off by Gaudreau's personality and aggressive nature. "Marcel told me that what he wanted to do was pack up Stan, his computers, his cell, and take the whole thing back to MIT," Rossi said. "It was clear that none of that was going to happen. So then what he wanted to do was just come and work as one of the people in Stan's lab. I didn't think that was so outrageous. I tried to promote that idea to Stan and see how he felt, but he had no intention of allowing it."

Finally Pons appeared and talked to Brophy in private, leaving the two men in the nice suits standing outside with Gaudreau. Then Brophy reappeared and departed with the two suits, leaving Pons to tell Gaudreau that he couldn't see him after all. Something had happened, Pons said. He didn't say what. He did promise that he would send a copy of the paper to MIT at the end of the week. Pons also said that the cold fusion work was eighteen months premature, but Gaudreau didn't ask him what he meant by that.

"You know, Dr. Pons," Gaudreau said, "I know you're being pressured to come up with these neutrons, but if you have fusion without neutrons, it'd be a hell of a lot better thing. These neutrons are going to cause you grief."

To which Pons replied, as best Gaudreau could remember, "If they want neutrons, I can make 'em neutrons."

To Stan Pons, Gaudreau must have looked like Houdini, appearing and disappearing outside his lab throughout the week. Tuesday he was there, although he did nothing more than exchange pleasantries with Pons. He spent the next two days in Oak Ridge, Tennessee, where he had a scheduled course on remote robot manipulation at the Oak Ridge National Laboratories. Friday and Saturday he was back outside Pons's door. After each brief contact with Pons, Gaudreau would run to a pay phone, call Ron Parker, and fill him in. Then he would hustle back, hoping to find Pons somehow more accessible than before.

Gaudreau said he never associated with the many reporters loitering outside the lab. If he saw any cameras or trench coats, he'd take the elevator down a floor, then shoot through one of the long corridors and try to approach Pons's lab from a different direction. Or he would wait until the corridor was empty. The U had posted two security guards by the elevators outside the lab, and if one or the other wasn't around, Gaudreau would sit in their chair. That way if someone asked him what he was doing, he could say that he was waiting for Dr. Pons, which was true. "I was always very, very careful," Gaudreau said. "I mean, I really thought that I was MIT's representative."

Gaudreau left Utah one week after he arrived, when one of the senior members of the Utah chemistry department called Mark Wrighton, who called Parker and suggested they get Gaudreau out of there before he got in trouble. At one point the University of Utah police sent a message to Pons's lab referring to Gaudreau as the guy "who Stan can't get rid of" and suggested that the next time he appeared, someone sneak away and call security, which will send someone out to "deal with the nuisance." It's possible that by the time this message was sent, Gaudreau was already on the plane home.

10 THE PRESS

On the morning of March 27, *The Wall Street Journal* reported Pons and Fleischmann's rationalization for the disturbing lack of neutrons. As the *Journal* explained, "Fusion physicists are accustomed to thinking of fusion reactions occurring in fractions of a second at enormously high temperatures and densities of hydrogen atoms. The reactions created by the two chemists take place over hours inside a solid crystal."

Thus, Pons said, "There's no reason the reaction [in the palladium] has to be the same" as that seen in the run-of-the-mill fusion reactions with which nuclear physicists are familiar. From this evolved the working hypothesis for cold fusion that something almost magical happened to the fusion process within the cold molecular lattice of the palladium.

This theory seems to stand in contradiction to one of the basic tenets of science, which is that the laws of nature here are the same as the laws of nature elsewhere. The laws of conservation of energy and momentum, for example, apply equally on the moons of Jupiter, in the core of a neutron star or a red giant or a supernova, or on the Lexington Avenue subway at rush hour. Science traditionally progresses by expanding the consequences of those laws known to be true in familiar realms to those realms into which no one has yet been able to look.

If the laws of physics were different inside a piece of palladium under electrolysis, it would open a Pandora's box, at least as far as the pursuit of knowledge went. What if they were only different in some pieces of palladium and not in others? Or what if every piece had a different variation on a theme? Or maybe the laws only differed in one piece of palladium? Mine, for instance. This sounds absurd, but consider that representatives of Johnson Matthey, the firm that supplied Pons and Fleischmann with palladium, took to talking of one particular preparation of the metal as "fusion grade." Indeed, it could begin to sound suspiciously like magic. But once this whimsical notion was accepted as plausible, that remarkable and unique properties existed within the crystalline lattice of palladium, anything was possible.

It may have been for this reason alone that the scientific community was soon split into believers and disbelievers. A belief in cold fusion required an act of faith from the start. And faith, traditionally, has had no place in science, where unbridled skepticism is considered a virtue. So some had faith and became believers, or "true believers," as they were sometimes called, an appellation that was not intended to be complimentary. To believe in fusion without neutrons was to believe in a benevolent universe, or a benevolent God. Neither, traditionally, had ever been so kind.

11 CAMBRIDGE

Thanks to Marcel Gaudreau, the MIT scientists knew on Monday, March 27, that the Pons-Fleischmann electrolyte was lithium deuteroxide. "That was like getting the first commandment," Gaudreau said later, "when you had no commandments at all."[9] And they knew that the palladium rods were the size of a pencil, which gave them an idea of the scale of the apparatus. And they knew that the rods were supplied by a company called Johnson Matthey, which gave them three commandments by Gaudreau's count.

Dave Albagli and Martin Schloh called a Johnson Matthey subsidiary in New Hampshire and were told that the firm had only two six-millimeter palladium rods in stock. So Mark Wrighton's chemists weren't the only parties who had become interested in palladium since the March 23 announcement. "Hold them," Schloh said. "Package them. We will be there right away." Albagli and Schloh must have had an overpowering sense of urgency, because they got a speeding ticket on the drive. Still, when they arrived they were informed that they could get only one rod; the second had already been purchased by a quicker customer.

Meanwhile, Mark Wrighton met with Ron Parker and Paul Linsay, a former particle physicist who had recently signed on with Parker's fusion group. Vince Cammarata and Dick Crooks from his own group rounded out the assembly. Wrighton said he was willing to do whatever was necessary to pursue cold fusion. His personal opinion seemed to be one of scientific impartiality. Wrighton believed that if cold fusion were true, "this revolutionizes everything and we are going to have to take a serious look." And if they couldn't be the first to discover, they could at least be the first to confirm. Wrighton offered the continued expertise of his chemists in return for the know-how of Parker's physicists. Parker took him up on it.

Wrighton had copies of the morning's *Wall Street Journal* article on cold fusion, which included a warning from Stan Pons that these fusion experiments could be extremely dangerous:

> He said that in an early stage of the experiments the apparatus suddenly heated up to an estimated 5,000 degrees, destroying a laboratory hood and burning a four-inch-deep hole in the concrete floor.

Even Wrighton seemed amazed by the violence of this episode. Linsay, the physicist, asked him if he knew of any chemical reaction that could do that. Wrighton said no. So the physicists, who certainly had their doubts about the plausibility of cold fusion in terms of nuclear physics, were temporarily mollified. If the chairman of the MIT chemistry department couldn't imagine how this explosion could come about, then maybe cold fusion had more validity than they thought.

Monday night the Parker-Wrighton collaboration set up five cold fusion cells at the Plasma Fusion Center, in a room that had been built to house MIT's next generation of tokamak, called ALCATOR-C MOD. This reactor was expected to generate kilowatts of power from deuterium-deuterium fusion. That was not much power by the standards of a modern-day electricity generating plant, but it did represent a very lethal dose of neutron radiation. Hence, the room that would contain ALCATOR-C MOD was a cube, thirty feet on a side, with five-foot-thick concrete walls to absorb the neutron bombardment. Portholes in the bunker allowed cables to run from the tokamak up to a control room that looked down on the cube from an angle.

The five cold fusion cells were put on a table in this bunker. Thermometers[10] were placed in the cells. Neutron and gamma ray detectors surrounded the cells. The researchers then set up their makeshift control station on the safe side of the portholes. This setup was what the MIT people referred to as their stage one calorimetry, a crude method of measuring the heat released by the cells. It would register fusion reactions if they were as dramatic as Pons and Fleischmann had claimed. It would probably not be sensitive enough to detect reactions that were considerably less dramatic.

The chemists calculated that it would take roughly thirty hours to saturate their palladium electrodes with deuterium. So they set the cells charging and took notes. None of them had ever done calorimetry before—not the electrochemists or the physicists. They assumed that if the cells began generating watts of heat, their various instruments would show it. Stan Luckhardt, one of Parker's physicists, said, "We were just going to sit back and watch the temperature rise."

That this image has its absurd aspects did not escape notice by the participants. Not the least of the absurdities was the sight of this multimillion-dollar building, awaiting its $100 million hot fusion reactor. There, bubbling away on a table in this barren bunker, was perhaps a thousand dollars' worth of chemistry equipment that had been advertised to do the very same thing. And sitting outside were a handful of chemists and physicists staring through these portholes, rubbing their eyes and waiting for salvation. They were skeptical but excited nonetheless. Like agnostics at Lourdes, they didn't particularly believe in miracles but wouldn't have minded seeing one if the opportunity arose.

Then there was the group from the materials science (metallurgy) department at MIT, which also wanted to do cold fusion experiments and had apparently been pointed in the direction of the PFC by the radiation safety office. They came by with file folders tucked under their arms bursting with *Wall Street Journal* articles and faxes. It was obvious that they had all the same information that Wrighton and Parker's people did, with the exception of what Gaudreau had gathered out in Utah that morning. "They were just dying to get their hands on what was in the electrolyte," said Dave Albagli. Wrighton's chemists and Parker's physicists wouldn't tell them. Eventually, these interlopers were absorbed into the collaboration.

The materials scientists were led by Ron Ballinger, who had a joint appointment in nuclear engineering and materials research. For the chemists and physicists, Ballinger's point of view turned out to be worth hearing. Apparently the two groups got to talking about Pons's mysterious explosion/meltdown. Ballinger said he wasn't as impressed as the physicists and chemists were. So what if they had an explosion? he asked.

Ballinger was a very talkative guy, a natural raconteur, and he had all kinds of stories about palladium. He told his audience that palladium is used as a catalyst in wood stoves, for instance, or the catalytic converters in automobiles. It is a surface on which chemical reactions will cheerfully take place. He said when you throw a palladium catalyst into a wood stove, you get what are called exothermic recombination reactions, and they can raise the temperature 300 to 400 degrees centigrade in a minute. If palladium happens to be in a hydride state, which is to say the hydrogen has bonded to the palladium molecules, and the palladium starts to erode and the hydrogen leaks out, this hydrogen could recombine with the oxygen in the air. It would be no problem whatsoever to make one of these electrodes glow red hot and melt. And he said that when a palladium rod becomes saturated with hydrogen, it expands. The hydrogen begins to leak out and recombines with the oxygen. And it does so on the surface of the palladium. Heat is released. The hotter it gets, the faster the hydrogen comes out, the more heat is released. And it has nowhere to go. Bang! Or the level of the electrolyte in the cell can drop, exposing the rod to the air. Bang!

Hydrogen-oxygen recombination happens to be an extremely powerful chemical reaction. Space shuttles, for example, are launched into orbit when their rocket engines recombine hydrogen and oxygen. This reaction, Ballinger explained to his audience, has one of the highest energy densities of any possible chemical reaction, which is to say it packs a bigger wallop than anything else you could mix in a laboratory. This is a rocket fuel, he said. What you need for a rocket is a big bang for your buck. (In fact, the potency of the effect in palladium had been recognized for 160-odd years, and it had even been used productively in cigarette lighters.) So maybe the mysterious explosion that Pons had mentioned that morning in *The Wall Street Journal* was not so mysterious.

While Ballinger talked, the MIT cold fusion cells continued to charge. And as the palladium absorbed the deuterium, it began to expand. The heavy water began to evaporate. The rods began to blacken, fatten, and crack. After a while the scientists pulled out one rod that had been only partway submerged. "It looked," said Paul Linsay, "like one of these ten-thousand-year-old fertility balls."

12 HARWELL

BRITISH BOFFINS BID FOR NUCLEAR BLAZE OF GLORY
March 25, headline Liverpool *Daily Post*
ATOM TEST MAN SEEKS SUN SECRET
March 28, headline Northampton *Chronicle & Echo*

The Harwell laboratory is situated about fifty miles west of London on what used to be a Royal Air Force base in the idyllic British countryside. From the outside it looks like a sprawling industrial complex or maybe an architecturally uninspired institution for higher learning. However, visitors must be cleared through a guardhouse at the front gate before entering the property. These security measures are necessary because the better part of Harwell's resources are dedicated to nuclear research. In 1986 the British government finished the transformation of Harwell into a commercial laboratory, funded through its customers in both industry and business, with the honorable goal of making money. Cold fusion, if it worked, would do just that.

Martin Fleischmann had spent the long Easter weekend at home with his artist-wife in Tisbury, then announced on a radio interview that he would be at Harwell Tuesday morning to give a seminar. This prompted a plague of reporters to descend on Harwell, only to find that it was closed for the weekend and anyone who was in working wasn't talking. At one point an Italian crew threatened to land a helicopter in the research director's backyard if he wouldn't give them an exclusive interview. The research director, Ron Bullough, apparently convinced them that an aerial assault would not work to their favor.

For the most part, the British press had been marginally less optimistic about cold fusion than their new-world counterparts. Although the reporters had minimal access to the scientists at Harwell, these already had ten days' experience in cold fusion and could curb enthusiasm. "It is not fair to encourage ridiculous optimism," Bullough told the reporters, and for the most part they did not. Still, that the Harwell staff knew about cold fusion and was taking it seriously was evidence that the work might be credible.

The resident scientists at Harwell assigned to the cold fusion beat were as curious as the media about what Fleischmann would have to say on Tuesday. Maybe more so. They had spent the weekend running cells and losing sleep. They had observed no neutrons coming from the cells and had spent Good Friday constructing more. The administration gave them carte blanche to take what they needed from the Harwell storerooms. With police assistance, they went around the stockrooms with a trolley helping themselves to anything they thought they might need.

By midnight on Good Friday, half a dozen cells were up and running, with neutron detectors and gamma ray detectors, and all the electrochemistry under computer control. The bulk of the work had been done by four scientists: two physicists, Martin Sené and David Findlay, and two electrochemists, Dave Williams and Derek Craston. These four insisted that they weren't overly anxious about the radiation hazard— "Well," said Sené, "we knew Fleischmann was alive"—but the director of Harwell was, and it was his laboratory. As Williams put it, the director had become "amazingly twitchy" about radiation. So they covered the cold fusion cells in plastic boxes, a precautionary measure should radioactive liquid suddenly squirt from the cells. They set up trip systems that would immediately shut the cells down should neutron and gamma rays begin radiating outward. They installed closed circuit television cameras to watch the meters on the radiation detectors, and those pictures were fed back to a nearby accelerator control room, which was manned twenty-four hours a day. The accelerator operators were then given a list of phone numbers and instructions: if the meters go above a certain level, phone these numbers and yell for help.

Dave Williams was a longtime friend and collaborator of Martin Fleischmann, and he believed that Fleischmann was close to a genius, "a fantastically creative guy." If he said something worked, it most likely did work, even if other experts said it didn't. Williams labored away at cold fusion cells until two in the morning on Saturday the twenty-fifth. Reporters woke him at eight in the morning. As he remembered it, "Some guy from the *Daily Mirror* wanted me to say that everyone was going to have a little [cold fusion] heater in the corner of their living room, and that this is going to provide all the energy needs for the nation."

When Fleischmann arrived at Harwell on the morning of the twenty-eighth, the day before his sixty-second birthday, he was confronted with an impressive gathering of the more expert researchers at the lab. His audience also included distinguished scientists from the UK's Science Research Council, Central Electricity Generating Board, Atomic Energy Authority, and the huge magnetic confinement fusion program known as JET, for Joint European Torus, located nearby at the Culham Laboratory. "I have never seen so many FRS's [Fellows of the Royal Society] gathered in one room in my life," said Findlay.

Fleischmann began his seminar with his heat data, or the "heat thing" as Williams later called it. Fleischmann flashed a series of tables and charts that seemed to demonstrate that the amount of heat emitted by the fusion cells was greater than anything they could explain by a chemical reaction. Fleischmann's learned audience seemed to find the figures for heat generation impressive. Still, Fleischmann was asked if he had done control experiments with light water, and he responded that they hadn't had time because of the frantic rush to go public.[11] Ron Bullough wondered how Pons and Fleischmann could have supposed they had made a discovery if they hadn't done any controls. With that, Bullough said, "he went down about five notches in my view."

Fleischmann turned next to his radiation data, beginning with what is known in the business as a gamma ray spectrum. These gamma rays were indirect evidence for the existence of neutrons but the primary evidence for fusion, as far as physicists were concerned. If the cell was emitting neutrons, then the neutrons would interact with the water in the surrounding bath, and that interaction would cause gamma rays to be emitted. So Fleischmann showed his gamma ray spectrum, and Peter Iredale, who was director of the Harwell Laboratory, looked at Fleischmann and said, "It's wrong." Then the other physicists in the room echoed his comment. To them it didn't look like a gamma ray spectrum, and they knew what a gamma ray spectrum should look like.

How Fleischmann felt about the pithy and unequivocal Harwell criticism can only be guessed. He said that he would call Stan Pons back in Salt Lake City and ask him about the spectrum. Derek Craston, who was there, recalled that even with a weekend of rest, "Fleischmann looked like a very tired man."

Still, the gamma ray spectrum was not Pons and Fleischmann's only evidence of fusion. The two had plenty more, especially the heat that had been generated, and the Harwell scientists had been impressed with that. And Fleischmann was still anxious to have the Harwell scientists reproduce his results. After the seminar, he spent an hour with Williams and Craston, describing how to do the cold fusion experiment. He even sketched a detailed schematic diagram of a fusion cell, to make absolutely certain that they would replicate his experiment correctly. That afternoon Fleischmann spoke with reporters and was remarkably frank. Cold fusion, he said, represented either "a Nobel Prize or a lot of egg on my face."

Fleischmann seemed to be momentarily pessimistic. Cold fusion, he admitted, could be nothing more than a "horrible chain of misinterpretations and accidents."

13 SALT LAKE CITY. GROUND ZERO

On March 28, Marvin Hawkins was told that the police were after him. The twenty-seven-year-old graduate student, who would refer to cold fusion as "my baby" and the fusion cells as his "little puppies," had been having one shock after another since his adviser and his university had decided to take cold fusion public.

As Hawkins told it, two days before the announcement he learned that his name would not be on the cold fusion papers, nor would he be permitted to participate in the press conference. After doing at least a good part of the lab work on cold fusion, Hawkins had reason to expect he would be listed as an author on the seminal paper, which would have meant a share of the Nobel Prize and any royalties. Now his contribution would not even be acknowledged.

"Stan said that the university had decided that I had to play a lower-key role in the announcement," Hawkins explained. "The lower-key situation was that I wasn't going to be involved in the press release and that they decided not to include my name on the paper. In that stage of the game you're kind of going, 'Oh well, okay.' And then I was walking across the campus and it just dawned on me like a brick that they had written me out. That, if it was the university behind it, the university simply said this man isn't high enough profile. He's not good enough to be here. And indeed that's exactly what happened."

Hard as this had been for him to accept, Hawkins convinced himself that he should simply shrug it off. "I said, 'Okay, I can deal with this.'" Hawkins was a good soldier. He even accepted his exclusion from the press conference, because he did not want to "throw any muck" on such an important occasion.

Then there was the matter of the missing lab books, which Hawkins said he put in his brother's safety deposit box. Pons's version of the story, as he told it to Hugo Rossi among others, was that Hawkins decided to give the books to the Church of Jesus Christ of Latter-day Saints as the modern equivalent of Joseph Smith's tablets, maybe because Hawkins was disgruntled by his exclusion from posterity. In some versions of the story, Pons said Hawkins tried to sell the books to the Mormon church for a million dollars. After a week of frustration trying to retrieve the books, Pons got a call from the LDS headquarters, saying the books were there and would he please come down and claim them. Rossi tended to believe this version, but then Rossi at this time tended to believe Pons. The bulk of the evidence supports Hawkins's version.

In any case, Tuesday, March 28, Pons arrived at the laboratory to find only the photocopies. Hawkins was still in the mountains skiing with his family. He said he called Pons first thing in the morning to check in. "I had called immediately around eight-thirty, and [Pons] said, 'But look, I've got to have these notebooks, the original data, and I have to have those for the patents,' and he's going on and on about this. And I said, 'Good grief, I didn't realize that it was that big of a thing.' I said, 'I'll come down right now, and I'll drive to my brother's, and I'll find out.' But I didn't have my brother's phone number where he worked. . . . So we come down from Park City, my wife and I, and we go into Stan's office, and he's just absolutely livid. He said that there's a warrant out for my arrest and if I had any sense about me I would get those books just right now, and he went on and on. I'm going, 'Good grief! Get real!' My wife is just in tears. This guy's been threatening me. He's going to put me in jail!"

Hawkins spent a few hours finding a lawyer, who told him Pons had no case and added, "Please, I beg of you. Please let them arrest you. We'll both be wealthy the rest of our lives." The lawyer, according to Hawkins, told him to keep the lab books as long as Pons had been given the photocopies, but Hawkins observed that "that sounds like a lawyer." His wife told him to "be careful, or you'll get burned," but Hawkins insisted Pons had been good to him, and he returned the books.

"Now we're in fine shape," Pons supposedly told Hawkins.

Hawkins then went down to the lab to find that he had been locked out. Pons had replaced him on the cold fusion project with Mark Anderson. Then Pons had instructed Anderson to remove all the palladium from Hawkins's desk and have the locks changed on the laboratory door to keep Hawkins out. Anderson did both.[12] Hawkins returned, infuriated, to Park City. Anderson then took to calling Hawkins for advice. "So he starts making four or five calls to me every day," Hawkins explained. " 'How do you do this? How do you that?' And I was hurt, but I figured it's still a bunch of bull."

14 AUSTIN

After working for four days to duplicate Pons and Fleischmann's experiment, Al Bard was getting leery. He had already decided to put together a more professional experiment, so he and Norman Schmidt had built a suitable electrolysis cell. They still, however, had little idea of what conditions were necessary to induce fusion. They had electrolyzed away, calculating how much heat was coming out of the cell and how much was going in. The accuracy wasn't that good. The experiment would only detect large amounts of excess heat. Still, the watts out were always fewer than the watts in.

By Tuesday, March 28, Bard was trying to find neutron detectors. He called his old friend Doug Bennion, the BYU electrochemist. Bennion told him that Steve Jones, the BYU physicist, had also done the cold fusion experiment and had detected neutrons emitted. So now Bard knew that someone else had confirmed Pons and Fleischmann, or so it seemed. Bennion said that Jones was using lithium deuteroxide for the electrolyte, just as Pons and Fleischmann had.

Then, surprisingly, Bard heard from Stan Pons himself. The American Chemical Society had asked Bard to give an introductory talk at a special cold fusion session the society was throwing at their spring meeting in Dallas on April 12. Bard said he'd do it, but he wouldn't talk in a vacuum. He had to talk to Pons and find out "what the heck he's really done." The ACS reached Pons and told him to call Bard, which he did, and he gave Bard the details. He told him everything. It may have helped that by now Pons had heard from Fleischmann about the Harwell critique and knew they needed support.

Pons told Bard that they had used a one-millimeter palladium rod, ten centimeters long, that they had set this within a cage of glass rods and wound a platinum wire around that. "Spacing is not critical," Pons said, "but use about half a centimeter. Run it at 200 to 300 milliamps for 7.7 hours to initiate fusion." Pons also told Bard that Jones had measured neutrons, confirming their work. And he told Bard about the meltdown. "So be careful," Pons said.

Bard and Schmidt went back to the laboratory and began the cold fusion experiments once again, this time using the recipe prescribed by Pons himself. The new experiments, however, failed to produce signs of fusion.

15 LOS ALAMOS, NEW MEXICO

Immediately after the Utah announcement, the Los Alamos National Laboratory spontaneously blossomed with unauthorized cold fusion research. It started with two experiments, but soon a dozen teams of scientists were trying to duplicate cold fusion. The lab had been founded by the U.S. government in 1943 in the canyons and mesas above Santa Fe as the center of the Manhattan Project atomic bomb research. Since then it had evolved into forty-three square miles of national laboratory with a budget pushing a billion dollars a year. Some $60 million of this was spent annually on two conventional fusion programs.

By March 28, the Los Alamos administration had officially appointed Rulon Linford to head up the lab's cold fusion effort. Linford was director of the magnetic fusion energy program and head of the controlled thermonuclear research division. Coincidentally, he had obtained his bachelor's in physics in 1966 at the University of Utah, where his father had once been head of the physics department. He had gotten his doctorate at MIT.

With Linford coordinating cold fusion at Los Alamos, the contacts between the laboratory and Utah began to grow like spiderwebs. On the twenty-eighth, Linford heard from the governor's office in Utah, which wanted to know what Los Alamos thought of cold fusion. Then he got a call from Ryszard Gajewski at the Department of Energy. Gajewski said he thought that Jones was right, and he had decided to fund Pons and Fleischmann, who had submitted a proposal to his office. Gajewski encouraged Linford to make contact with the University of Utah and see what he could do to help straighten things out.

Meanwhile, Pons had his own contacts in the physical chemistry division at DOE, and they had already put him in contact with Robert Sherman, a physical chemist in the tritium division. It may have been through Sherman that Pons gathered the mistaken impression that Los Alamos had successfully reproduced his experiment, which is what he told the local papers. (U. CHEMIST BELIEVES FUSION CONFIRMED AT LOS ALAMOS, reported the March 28 *Salt Lake Tribune,* adding that Los Alamos officials would not confirm the report.)

Tuesday afternoon, Linford called Pons and miraculously got through. Pons said that the powers-that-be in Utah were arguing against a collaboration with Los Alamos. According to Linford, "He articulated to me that these people wanted to obtain as much control over this discovery and as much economic benefit [as they could], and to keep it within the state. They didn't want a federal lab to take it away from them and run with it. [Pons's] argument for wanting this contact was if it really was nuclear fusion, there may have been defense- or weapons-related spin-offs, or developments that could come from this that would clearly have not been controllable by the state of Utah. The government would step in anyway. It was better to have a lab like Los Alamos, which has all the knowledge and expertise, become involved and make that assessment quickly."

Linford asked Pons what the Utah chemist would like from Los Alamos, and Pons told him they needed help to understand exactly what it was they had discovered.

On Wednesday, the twenty-ninth, Linford received an even stranger call, this one from an assistant secretary for security affairs at the Department of Energy. It seems this DOE fellow was "deeply apprehensive" about the weapons-related capabilities of cold fusion and wanted an informed opinion from Los Alamos. In particular, he wanted to know if the federal government should step in and classify the research.

It seems that if cold fusion à la Pons and Fleischmann would emit copious neutron radiation, it would not be, in the words of one Caltech graduate student, the benign little gizmo that it appeared to be. It would be a terrifying device, an ideal way of making plutonium in one's basement. All that would be necessary would be ordinary uranium, which could be bought without too much difficulty. The neutrons would gradually convert it to bomb-grade plutonium, so for a few thousand dollars' worth of heavy water, palladium, and platinum, anyone with passable expertise would be a long way toward building a nuclear bomb. Fortunately, the press hadn't immediately noticed this aspect of the device, so it hadn't made the papers. But it was there all the same.

Linford had thought about this too. He told the assistant secretary that it was a little late to shut the barn door, as the technology had already been made public. Linford suggested that if federal agents were to swoop down and declare all cold fusion research top secret while the technology was being hailed by the press as salvation, the act would have nasty and explosive repercussions of its own.

16 SALT LAKE CITY. PHYSICISTS

For six days now, physicists at the University of Utah had been receiving calls from reporters and scientists who had made the natural but incorrect assumption that any discovery in nuclear fusion would have come from the physics department, or at least that these physicists would have been consulted along the way. All the physicists could say was no, they knew no more about cold fusion than anyone else did, which was nothing; try calling Stan Pons. Several of the physicists had tried to see Pons as soon as he returned to work on Monday but with no success. Finally, Jim Brophy arranged for a delegation of physicists to get access to Pons.

At eight in the morning on March 29, Craig Taylor, the head of the physics department, Mike Salamon, Gene Loh, and Orest Symko met with Brophy, who allowed them to study a copy of the paper that Pons and Fleischmann had submitted to the *Journal of Electroanalytical Chemistry*. Then the delegation walked over to the chemistry building and met with Pons, whom they found, to their pleasant surprise, warm, sensitive, and patient. "We spent an hour with him," Salamon said. "He seemed like a very reasonable guy. Levelheaded." Pons must have been making an effort, however, because Salamon and his colleagues politely pointed out that his evidence for cold fusion seemed to have potentially fatal loopholes. First there was the gamma ray spectrum. Salamon told Pons the same thing the Harwell physicists had said to Fleischmann—"It's wrong." Pons was not surprised by the news. "He was ready for it when we told him," Salamon said.

The entire discussion continued along these disheartening lines. Frankly, the physicists told Pons, as far as the evidence for nuclear by-products—for neutrons and gamma rays—was concerned, they felt there were absolutely no data.[13]

17 PROVO

After March 23, the various administrators and scientists at BYU who had been involved with the cold fusion negotiations felt, as Lamond Tullis, associate academic vice president, put it, that they had been "had." Tullis said they were overcome by "righteous indignation" and an "enormous amount of stunned incredulity." Jae Ballif, the provost, would say he felt "devastated that, for whatever reason, agreements between honorable people were not kept." Only Steve Jones professed to be relieved: now he no longer had to wonder what Pons and Fleischmann were up to. He knew, and it was done.

What did move Jones to righteous indignation was the continuation of the "rumors" that he had engaged in scientific larceny. Jones heard from a friend at Los Alamos that Bob Sherman, who had been in touch with Pons, was saying that Pons was saying that Jones had stolen their idea. He may have felt some stunned incredulity as well. Jones said, "I called Bob, and I said, 'What's going on here? If you have any questions, let me know. I'll be glad to answer them.' I told him what we had done. And he said, 'I'm sorry.' And that was the end of it."

Jones no longer professed to have doubts about the morality of his actions. He had been victimized by the press conference, and that was that. Pons, Fleischmann, and Peterson were, to use the lingo of law enforcement, the perpetrators. And Jones's colleagues at BYU believed they had no reason to doubt Jones, because they too had been victimized by the press conference. Although they professed to be absolutely flummoxed by the duplicity of Chase Peterson, a man of previously unimpeachable integrity, they still were not so perplexed that they would question the credibility of Steve Jones. They just could not figure out how to reconcile this paradox. It was one of life's mysteries: "I suppose [Pons, Fleischmann, Peterson] feel justified in what they did," said Tullis, "because they apparently acted on the assumption that we stole their technology, which was a flagrantly ill-informed position. It doesn't justify a total collapse in integrity."

On the day before the announcement, Jones began compiling his defense against future recurrences of the accusations. This was the official BYU history of cold fusion. It was Ballif's idea, and Jones provided the necessary facts. Ballif said he couldn't verify all the information Jones provided, but "I have no reason to doubt him." This faith was further encouraged by Ryszard Gajewski at the Department of Energy, who spoke to the BYU administration the day after the news conference. Gajewski said he and his colleagues were equally appalled by the turn of events. "His advice to us," said John Lamb, BYU's director of research administration, "was to try to stay above the noise and the confusion, to continue our scientific work, to use the normal scientific procedures." This was their inclination in any case, but it was nice to hear it echoed by DOE.

Meanwhile, Paul Richards, who was head of the BYU public communications office, had to sit back and watch the media buy without question what Pons, Fleischmann, and company were selling. "Every day [there was] some new announcement," Richards said, "and the press was falling for it hook, line, and sinker, and not confirming anything."

When the BYU work was mentioned, it was invariably taken as confirmation of Pons and Fleischmann. This conception, oddly enough, was propagated further by Gajewski. Reporters who called DOE for an official comment on cold fusion were directed to Gajewski, who said that he believed that fusion had been observed in a solid and that this was a "major scientific discovery." Gajewski, of course, was referring to Jones's work, but he would add that Pons and Fleischmann's heat output was inexplicable, and he had decided to support the Utah research to the tune of $322,000. This seemed like a wholehearted endorsement of the Pons-Fleischmann product by any standards.[14]

Until March 29, all Jones would tell reporters was that his work was independent and did not confirm the Utah results "in any way." Jones would insist that he wanted to do this right and wait for publication before talking about his data. (He would, however, admit that the discrepancy between his neutrons and Pons and Fleischmann's heat was a factor of a trillion.) All Richards could do was read callers the abstract that Jones had written for the American Physical Society meeting, which did not make for a satisfying story. Richards spent his time begging Jones to go public while fending off journalists who "were sniffing out that there was something fishy going on." Then *The Wall Street Journal* arrived on the morning of the twenty-ninth and, said Richards, "really called a spade a spade."

SECOND FUSION DISCOVERY COMES TO LIGHT, read the headline. FINDINGS AT BRIGHAM YOUNG ARE SHROUDED IN SECRECY, SEEM LESS CONTROVERSIAL. The story read in part:

> The Brigham Young scientists are refusing to talk about their findings until their report is published in a scientific journal, thereby avoiding the anger among researchers that the Salt Lake City scientists stirred up by announcing their discovery at a news conference Thursday.

The *Journal* also remarked that Johann Rafelski was developing a theory to explain the BYU results, which he hoped to "finalize" soon and submit to a scientific journal. Thus, Rafelski's theory was "bolstering the credibility of the Brigham Young experiment," which surely made it one of the few times in Rafelski's career that his theoretical speculations had been cited as adding credibility to a subject.

Richards finally invited the local television reporters down to Provo on March 29, and the reporters from *The Deseret News* and *The Salt Lake Tribune* the following day. When the press arrived, they were given a demonstration of the experiment and, of course, shown the lab note-books. They were also shown Jones and Rafelski's two-year-old *Scientific American* article, which was fortuitously entitled "Cold Nuclear Fusion," seeming to support Jones's thesis that he was doing serious electrochemi-cally induced fusion work long before Pons and Fleischmann's proposal came his way. That the article was on muon-catalyzed fusion seemed to be irrelevant. To the layman, cold nuclear fusion was cold nuclear fusion.

Jones reiterated to the reporters that the BYU results were much less dramatic than the Utah results. Did they promise energy salvation? "Not by a long shot," Jones said, rolling his eyes. He then said that the odds that he and his colleagues had made a mistake were only "about one chance in two million," which made the BYU results seem only a modicum less incontestable than the Pons-Fleischmann variety.[15]

18 "OBSERVATION OF COLD NUCLEAR FUSION IN CONDENSED MATTER," BY S. E. JONES, *ET AL.*

On March 29, Steve Jones's *Nature* paper seemed to appear everywhere at once, thanks to the magic of the facsimile machine. Apparently Jones gave copies of the paper to half a dozen scientists, who then began photocopying and faxing it to friends and colleagues. Whoever was on the receiving end of the fax would make photocopies to be distributed to everyone in the neighborhood, then fax the article on to some new locale.

"Observation of Cold Nuclear Fusion in Condensed Matter," as Jones's paper was called, represented a distinct positive upturn in the course of cold fusion. "This is a real paper and not a press release—it shows data," wrote one awed researcher, who typed the good parts onto a computer bulletin board for those interested parties who had not been blessed by a fax.

But those who read the BYU paper carefully were struck by how small the effect was—two neutrons per hour. In fact, if it was, as *The Wall Street Journal* had put it, "likely to be more immediately acceptable to scientists" than the Utah results, it was only because the effect was so close to zero as to be believable. It's questionable whether the scientific community would have found Jones's paper of interest at all had it not appeared before Pons and Fleischmann's and thus given the information-starved community something to chew on.

Chemists who found the minuscule BYU effect almost irrelevant nonetheless found Jones's reporting of his electrolyte worthy of comment. It quickly became known as "the Mother Earth soup," which is to say it seemed to contain a dozen different metal salts, possibly more:

0.2 g amounts each: $FeSO_4 \cdot 7H_2O$, $NiCl_2 \cdot 6H_2O$, $PdCl_2$, $CaCO_3$, $Li_2SO_4 \cdot H_2O$, $NaSO_4 \cdot 10H_2O$, $CaH_4 (PO_4)_2 \cdot H_2O$, $TiOSO_4 \cdot H_2SO_4 \cdot 8H_2O$, and a very small amount of $AuCN$.

When this recipe was followed to the letter, the various ingredients would cross-react, oxidize, and reduce one another and either plate onto the electrodes, creating what one chemist referred to as "a mossy mess," or settle, sludgelike, to the bottom. The BYU research, said David Findlay of Harwell, "was probably not the divine last word in nuclear physics experiments, but you still felt a nagging doubt at the back of your mind that maybe there was some process going on there."

19 PASADENA

Nate Lewis and his colleagues received the Jones paper on March 29 and immediately set about trying to duplicate that work. The Caltech chemists, however, were utterly mystified by the Mother Earth soup. They would refer to the electrolyzed concoction that resulted as, variously, a "piss-green solution with gray slag at the bottom" or "brownish red shit with this much garbage on the bottom," or simply "a total mess."

So Lewis called Jones's lab and was told not to use all the metal salts in each cell but only one or a few at a time. John Gladysz, who was on sabbatical at Caltech, discussed this in one of his e-mail notes home to Utah. (He called these notes "reports from the mole.") He wrote:

Interestingly, that Jones paper is in deep disrepute. It reads well, but Nate has talked to Jones on the phone . . . there is an experiment that reads as though they got fusion in an eight component electrolysis mixture. What is really the case: they didn't enter the brew into their lab books, all they know about the most successful expts you see quoted in their ms is that they contain "two, perhaps three, out of a possible eight components."[16]

Upon receiving the paper, Lewis and company immediately ran one Mother Earth cell, which emitted no neutrons. Lewis then decided to opt for a technique akin to overkill. He calculated that, given the efficiency of the Caltech detector and the production of Jones's cells—two neutrons per hour from as many as four to eight cells run simultaneously—they would need roughly fifty similar cells, all running at once, to assure that they would detect a neutron signal over the background. "So I said, We're going to build fifty cells," Lewis later explained, "and they said, Ahhhrrrghh."

These researchers promptly dubbed what they were building the VLFA, for Very Large Fusion Array, which was a graduate student's idea of a pun, no insult intended, on the Very Large Array, a network of radio telescopes in the Southwest. "We put together this thing with fifty test tubes," said Mike Sailor, "got this large sheet of titanium, analyzed it quickly with a scanning tunneling microscope. It was pure titanium. Cut it into fifty electrodes, wired it up, and covered the wire with epoxy, so the wires wouldn't get corroded in solution. Then the problem was that we had no good counterelectrodes. We didn't have fifty wires of platinum or gold. We didn't have that kind of money. But Charlie Barnes had this personal, research stock of gold. He just grabbed this chunk of gold and said, Go for it."

They also debated sacrificing their dental fillings, not to mention Lewis's wedding ring, to the cause of cold fusion. They agreed that the fillings could go, but Lewis refused to part with the ring. "We had *some* priorities still left," he said.

In any case, Sailor, Reggie Penner, Mike Heben, and company found a metal roller and spent the night of the twenty-ninth rolling Barnes's chunk of gold into thin plates and slicing them into fifty electrodes. These they put in fifty tiny cold fusion cells; then they wired the entire array for fusion. They took it over to the Kellogg Radiation Lab and put it inside the cube.

Although Pons and Fleischmann had been less than willing to tell anyone exactly how long one had to wait for fusion, Jones had stated clearly that his neutrons appeared within the first thirty minutes of charging. So Lewis and company ran the fifty cells for several hours and saw nothing.[17]

Meanwhile, Lewis did receive an e-mail message from Pons in response to one of his many requests for information, although it was of dubious assistance. The note suggested that Lewis "should look for the effect first and the cause second," and that was about it. Lewis observed that the message was only slightly less abstruse than the choice of ingredients in the Mother Earth soup.

20 SALT LAKE CITY, PHYSICISTS AND ADMINISTRATORS

At three o'clock in the afternoon on March 30, a delegation of Utah physicists took their critique of the cold fusion data to Chase Peterson. This time the delegation included Craig Taylor, the chairman of the department, Mike Salamon, Hugo Rossi, dean of the College of Science, and Joe Taylor, academic vice president at the university.

Taylor, who was trained as a mathematician, later took credit for getting the physicists access to Peterson. He had become very pessimistic about cold fusion. He had found the press conference mildly embarrassing and, a few days later, was further dismayed when he finally learned exactly how big the purported discrepancy between heat output and neutrons was. "I didn't understand that," Taylor said, "until a few days later when I was talking to Brophy. I asked him what exactly is the nature of the discrepancy, and he said it was a factor of about one billion. I nearly fell out of my chair. I remember telling myself, 'We are in very deep shit.' "

At the meeting with Peterson, Salamon did much of the talking. He told Peterson that as far as conventional nuclear physics was concerned, cold fusion was all but impossible and then explained that Pons and Fleischmann had a disconcerting dearth of evidence.

Peterson asked whether it was conceivable that the experiment did create fusion but that the energy of the fusion was converted directly into heat and not into neutrons. That would explain why Pons and Fleischmann had observed so few, if any, of them. Craig Taylor (no relation to

Joe Taylor) explained the rudimentary physics to Peterson: what we perceive of heat is the result of the vibration of entire molecules. A fusion reaction, by contrast, is what happens when the nuclei of two atoms fuse together. These two mechanisms exist on hugely different scales of distance and time.

When fusion occurs, the newly formed nucleus will stabilize itself by ejecting a neutron or a proton in about 10^{-22} second (0.0000000000000000000001 seconds), which is blindingly fast. Atoms, by contrast, will vibrate about every 10^{-13} second. This is also quick by human standards, but it is one billion times slower than the decay of the nucleus, which is to say, the same ratio as thirty years to one second. As Taylor explained it to Peterson, the nucleus ejects this neutron with such instantaneous violence that everything else in the neighborhood appears to be virtually frozen in time. And the neutron, or whatever, is gone from the area well before the nearby molecules even know it existed. Expecting this particle's energy to be somehow transmitted into heat, said Taylor, would be like shooting a pea at the speed of light out into our solar system and expecting it to noticeably warm up the planet Jupiter as it passes by.

Under the circumstances, Salamon and Taylor argued, it was in the university's best interest to take a very conservative and low-profile position on cold fusion.

Peterson gave short shrift to the delegation's advice. Perhaps it was too little advice too late, or a case of asking Peterson to close the barn doors now that he'd personally let the horses loose. Peterson had met the day before with Governor Bangerter, as well as the board of regents and various legislative leaders, in what the papers had called a "closed door session." Bangerter emerged announcing that the state legislature would hold a special cold fusion session on April 7, at which point he would urge them to appropriate the $5 million for cold fusion research. Peterson had announced that James C. Fletcher, who had just tendered his resignation as head of NASA, had decided to return triumphantly to the University of Utah, where he had once been president, to direct the cold fusion program. (Fletcher was interviewed in Washington by *The Deseret News*. With frightening syntax, he explained that cold fusion, if real, "would be the best thing that's happened since I guess the atomic bomb—and that was bad, but in terms of breakthroughs it would certainly be a bigger factor than the transistor when that was built." NASA then announced that Fletcher was not planning to run the Utah cold fusion effort and that Peterson, apparently, had been the victim of a misunderstanding.)

Peterson was not about to ask the governor to take back his $5 million offer. He did tell his physicists that he would couch the university's hopes for cold fusion in a more cautious and conservative tone. And, for a day or so, he did.

"I admit we could be wrong," Peterson told reporters. "The science is good. It [the fusion experiment] is being repeated, we're told, in labs in the U.S. and overseas. But there could be a hangup, a glitch, that we don't know about." To this statement, *The Deseret News* added: "It's possible a reactor could prove unsafe, for reasons not now foreseen, or that the energy produced could be so costly as to be economically infeasible. If the commercial research comes to such a roadblock, work will stop, Peterson promised, and no more money will be spent."

Jim Brophy also tried publicly to contain his enthusiasm after the physicists voiced their doubts. That same afternoon Brophy gave a memorable interview to *The Deseret News:* "If we had announced a breakthrough in energy-producing events," he said, "no one would have shown up [at the press conference]." He then added, "I don't mean to say we used fusion falsely. There is fusion going on in the platinum lattice." And then, while not admitting that they might be wrong, he did address the possible repercussions to the university's esteem in the field if they were: "But there may be a taint that we can't quite wash off. A graduate student may decide he'd rather go somewhere else or something, but there's nothing we can't handle."

What made this remark particularly poignant was that the rumor mills were already tying cold fusion to an earlier Utah scientific debacle. This was the X-ray laser affair of 1972. Although this had been a decade before the arrival of Stan Pons, the rumors spitefully connected him to that fiasco as well.

There were, indeed, some disquieting similarities between the X-ray laser and cold fusion. The first was the brainchild of John Kepros, a postdoc at the time, working under chemistry professor Edward Eyring.[18] Kepros and Eyring believed that by firing an ordinary infrared laser into a copper gel "sandwich" they could generate an X-ray laser out the other side. The press caught on to the discovery, and it made the papers, including such discerning journals as *Newsweek*. This prompted a circa 1972 delegation of Utah physicists to march on the administration and condemn the chemists for going public with a physics discovery that they seemed to understand not at all. In fact, no known physical effect could explain how such a gel could produce a coherent beam of X-ray light. It was soon demonstrated, however, that the data could be explained without invoking X-ray lasers.

The X-ray laser affair may have spawned the disparaging term *Utah effect,* which was thereafter applied freely to any public relations disaster originating within the state. (Cynics were already suggesting that cold fusion was the latest product of the Utah effect.) Still, the X-ray laser affair did not strike Peterson as a particularly portentous allegory. He later told the state legislature that after the X-ray laser fiasco the national ranking of Utah's chemistry department rose dramatically. Hence, he said, "we should not operate from the premise that [early action] will embarrass us."

21 THE NETWORK

Without a doubt, the dominant factor in the first week of the flowering field of cold fusion research was the lack of information. Martin Fleischmann estimated that in that time he had given details of his experiment over the telephone to several hundred scientists. Yet the available information was depressingly unreliable. Larry Faulkner of the University of Illinois, for one, estimated that in the first few weeks after the Utah announcement he spent some 30 percent of his time on the phone comparing cold fusion notes: "The consistency of what we were getting out of Utah was extremely poor," he said.

After the telephone and fax machines, the major source of cold fusion information was the various computer networks—in particular, Usenet and Bitnet—which link universities, industry, and government facilities throughout the world. These networks had begun carrying requests for cold fusion information within hours of the March 23 announcement. It was, as writer and physician Lewis Thomas would put it, a "collective derangement of minds in total disorder," played out on neurons of semiconductors. The net spread a combination of information, rumors, requests, and reports. The first *Financial Times* article, for instance, had been typed into the "net." It had hard facts, provided by Fleischmann and missing from other newspaper reports. And, of course, because Stan Pons had a computer and was connected through the network to the outside world, his electronic mail address had been disseminated on the net as well:

Here's Dr. Pons's e-address:
pons@chemistry.utah.edu
You can try to finger it first.

Two physicists, who requested to remain anonymous, did try to "finger" Pons's mail account. To assure that they wouldn't be traced, they used the account of another physicist who rarely used his computer. They said later that they were "sweating like thieves breaking into a bank vault," expecting the police. Pons's account was guarded by a password of his choosing. The two physicists tried "cfusion," which didn't work, then remembered that Pons had said he was a homebody, so they tried the names of several family members and hit eventually and correctly on Sheila, the name of his wife. So they broke into his computer account and spent two hours reading his mail. They were hoping for any clue: perhaps a letter to Fleischmann saying Pons had tried the palladium-tungsten mixture or some such and it worked, or maybe an order form to Johnson Matthey, the suppliers of the palladium.

They justified the breaking and entering by the importance of the controversy and the inability to get a straight story out of Utah. One of the two said they found "junk, nothing of interest, crap." He added that he knew of at least two other scientists who did the same, one from Oak Ridge and one from Caltech. In any case, Pons later changed his password.

Almost immediately after the Utah announcement, one could find on the network a variety of encyclopedic entries on deuterium and heavy water, as well as a lengthy compendium entitled "Everything you wanted to know about palladium and were afraid to ask," apparently cribbed from a book called *Guide to Uncommon Metals* by Eric N. Simons.[19]

The net also ran no shortage of theories to explain cold fusion should the experimental details pan out. This left an archaeological flowchart of the cold fusion speculation of at least one small segment of the scientific community. One entry, for instance, from Eugene Brooks, a physicist at the Lawrence Livermore Laboratory was entered at 00:45:36 GMT on March 26, which would make it midafternoon on Saturday in California. Brooks proposed a "solid-state process," in which deuterium atoms are sucked into the lattice of the palladium and maybe fuse by quantum mechanical tunneling.

Two hours later Paul Dietz, a computer scientist at the University of Rochester, signed on. He began, "Eugene Brooks proposed in a previous message a mechanism for how solid-state fusion might work," and explained *his* theory, which required that the palladium lattice be filled to the brim with deuterons. At that point, when one more deuteron entered, it would have to share a space somehow with another deuteron, or else, Dietz speculated, "pry open" the palladium lattice, an act of violence that seemed "energetically expensive." Thus, fusion might only be induced after all the holes in the palladium lattice are already filled.

Following Dietz by some eighty minutes was Matt Kennel of Princeton, who began "[Paul Dietz] writes: Eugene Brooks proposed . . ." Kennel then described his own variation, ending with a modest "How does that sound?"

> **Maybe it's possible that electrochemistry people have been having fusion all along with Pd experiments (in regular water, the rate would be lower) but nobody would be crazy enough to look for gamma rays, for god's sake!**

All three variations constituted what is disparagingly described as hand waving and required miracles of sorts to overcome the repulsion between the two positively charged deuterium nuclei, but such is the nature of speculation.

Kennel was followed by a University of Texas astronomer (14:41:49 GMT) who suggested, incorrectly as it turned out, that perhaps the reason Pons and Fleischmann didn't see neutrons was that their neutrons were absorbed in the palladium and heavy water. The Texan was then followed by a retransmission of a message from the twenty-fourth in which the correspondent reported, also incorrectly, that "the two guys who discovered the process are, together, the most respected electrochemists in the world." And so it went, twenty-four hours a day, every day, for months.

22 SALT LAKE CITY, GENEVA, NEW YORK CITY

March 31 promised a break in the general level of ignorance regarding who had done what with cold fusion, and how. Stan Pons was scheduled to give a cold fusion seminar at the U; Martin Fleischmann was booked into the main auditorium at CERN, the huge European physics laboratory outside Geneva; and Steve Jones was lecturing at Columbia University in New York.

All three scientists played to packed houses. The Utah seminar was held in the auditorium of the chemistry building, which seats approximately 400. Pons had not wanted the affair to become any more of a circus than necessary, so the administration agreed, after some negotiation, not to hold it in the university's basketball arena. It was estimated that a thousand people appeared hoping to hear Pons speak. Those who did get in had arrived at least forty-five minutes early. The local television crews were on hand, as well as NBC national news. Pons, however, would not allow video recordings, and the reporters had to settle for interviewing members of the audience as they filed out.[20]

At CERN, which is the preeminent physics laboratory in the world, a similar madness ensued. There, however, those who got into Fleischmann's seminar only had to arrive thirty minutes early. Then Carlo Rubbia, the lab's director general, refused to begin before ejecting the swarm of television and newspaper reporters. Rubbia explained that this was a scientific meeting, and he promised the reporters that a press conference would follow.

As for Jones, the organizers of his Columbia seminar had to relocate it three times before finding an auditorium that could handle the crowd. Still, it was standing room only. Amiya Sen, the Columbia physicist who had invited Jones, described the event in familiar terms: "Of course, when he arrived on campus," Sen said, "all hell broke loose. It was a circus."

At Utah, Pons said that they had produced twenty-six watts of heat for every cubic centimeter of electrode and that one cell produced four megajoules of heat in 120 hours.[21] All of this sounded awe-inspiring. He also admitted that they had seen a trillionfold too few neutrons to explain this mysterious heat and suggested therefore that the heat could not be coming entirely from deuterium-deuterium fusion, although no chemical phenomenon could explain this heat either. Pons said that the results were as puzzling to him as they were to everyone else and joked that they might see a viable commercial cold fusion technology in a hundred years. (In the early days of cold fusion, it was when Pons seemed to be joking that he seems to have been most honest.) One physicist standing in the rain after the seminar told reporters, "Physicists say that the excess heat must be due to chemistry, the chemists maintain that it must be nuclear physics."

Rulon Linford of Los Alamos had been invited by Jim Brophy to attend the seminar. Linford found the lecture puzzling more than revealing. "It was short," he said later, "and it was essentially philosophy: a table of results, of amounts, one picture of the flask in which the experiments were done, and some discussion of how the calorimeter was used. But there was never any raw data presented during the lecture. You came away with some idea of what he was claiming, but no understanding of some of the care, or amount of data they had, or any of those kinds of things that allowed you to really judge the quality of the work." Linford added, however, "Not being an electrochemist myself, I don't know what's usual in that field."

Fleischmann's seminar at CERN was little more informative but still seemed eminently scientific to the physicists present, few of whom were experts in either solid state or nuclear physics. Rubbia, for instance, who had his own history of announcing discoveries without sufficient supporting evidence, was asked his opinion, and said, "Dr. Fleischmann has planted a seed—will the seed grow up? I think yes."

At Columbia, Jones seemed more interested in the past than in the future. He spent most of his time discussing his muon-catalyzed fusion results, which were the original subject of the seminar, then spent most of the remaining minutes on cold fusion, showing photocopies of his three-year-old notebooks as evidence that he had not pirated the idea from Pons and Fleischmann. He did emphasize that his neutron rate was the equivalent of one trillionth of Pons and Fleischmann's heat emissions, and that Pons and Fleischmann's evidence for neutrons was meaningless. His cold fusion results, he said, did not confirm Pons and Fleischmann's research. He had observed neutrons from cold fusion; Pons and Fleischmann had not. The seminar was entitled "First Detection of Cold Fusion Neutrons," which said it all.

This became Jones's stock lecture, which he would repeat with minor variations in Sicily, Geneva, Santa Barbara, Baltimore, and Los Angeles in that order. In Santa Barbara, for instance, Steve Koonin of Caltech found his lecture "really disappointing. He spent the first half basically claim-staking and I found that really inappropriate." One observer in Los Angeles said, "Jones, to everybody's disgust, spent rather more time showing overheads of notarized pages in his lab book to establish precedence then he did actually talking about what he found in his experiments. That went over like a lead balloon. Everybody was saying, 'We don't give a shit who thought about it first. We want to know whether it works or not.' " It was the same story, but Jones did care about not only establishing precedence but distancing his results from those of Pons and Fleischmann. After all, he believed his were right.

Richard Garwin of IBM was at the Columbia seminar. Garwin had long before attained something akin to mythical standing in physics, as well as in the defense community. He had studied under Enrico Fermi at the University of Chicago. Fermi, who may have been the greatest physicist of the twentieth century, Einstein notwithstanding, called Garwin "the only true genius I've ever known." For his postdoc, Garwin built the first hydrogen bomb, apparently over his 1951 summer vacation. Hans Bethe, the Tolstoy of quantum mechanics, said of Garwin that he "is probably the smartest person I know," then noted that the only other person he knew with Garwin's level of talent was Fermi. William Happer of Princeton, who was director of JASON, called Garwin "a perfect experimentalist." He was said to have "exquisite common sense."

All of this made Garwin someone whose opinion one should weigh seriously, whether one agreed with it or not. Ryszard Gajewski had called Garwin just before the Utah announcement, informed him of the Utah-BYU contretemps, and asked him to be on a Department of Energy review panel that would assess the conflict. After March 23 that had become a moot issue, but Gajewski had sent Garwin the Jones preprint, and Garwin had decided to go to the seminar.

As far as the evidence for cold fusion went, said Garwin succinctly, "Jones didn't have enough data." Still, the audience in general, scientists and reporters alike, seemed to be less discriminating. "There were a few worried faces in the audience at the start of the talk," wrote one correspondent on the network; "at the end, the speaker received a raucous applause." *The Washington Post* reported that this was the first time "skeptical scientists and students" were presented with the results of a cold fusion experiment: "both appeared to be convinced that the fusion is real if not impressively large."

23 ''ELECTROCHEMICALLY INDUCED NUCLEAR FUSION OF DEUTERIUM,'' BY MARTIN FLEISCHMANN AND STANLEY PONS

It is possible to find the truth without controls, but the process has been demonstrated again and again to be notably

inefficient, so that years may be required before it is appreciated that a given treatment is worthless.

E. BRIGHT WILSON,
An Introduction to Scientific Research

The Pons-Fleischmann paper in the *Journal of Electroanalytical Chemistry* emerged from facsimile machines around the country just as the two chemists were giving their seminars. Ron Parker of MIT called it "a gift from God." The paper, of course, was copied and sent through the system of faxes, distributing His bounty far and wide.

The Wall Street Journal received a copy and announced the news in a headline the following Monday: HEAT SOURCE IN FUSION FIND MAY BE MYSTERY REACTION—FUSION PAPER SHEDS LIGHT ON FIND, EASING SOME SCIENTISTS' SKEPTICISM. The *Journal* reported that Pons had provided copies of his paper to five scientists, and from those five it had propagated prodigiously.[22]

In fact, the Utah paper, like the BYU paper two days before, was disappointing. It was, said Jim Brophy, "better than a telephone conversation, there's no question about that," which was certainly true. But that was about it. Al Bard said the paper read like a "rush job," and Dick Garwin called it "the worst so-called scientific paper" he had ever seen, adding that if the paper included the full extent of Pons and Fleischmann's data, the two chemists needed a miracle.

Pons, Peterson, and Brophy would later talk up the paper as though it had been peer-reviewed, but that had not been the case. Roger Parsons, the managing editor of the *Journal of Electroanalytical Chemistry*, had served as the sole referee. He later admitted cheerfully that he had accepted it because Fleischmann was his longtime friend and he respected him as a scientist. And the paper undeniably was newsworthy. "You take a chance," Parsons observed. "I suppose that I could say that I took too much of a chance on this one."[23]

The paper's most glaring and immediate deficiency was that it made no mention of any control experiments. This raised the pithy question of how Pons and Fleischmann could have come to any conclusion about their experiments, let alone such a remarkable one as room-temperature fusion. As E. Bright Wilson phrased it in *An Introduction to Scientific Research* thirty-seven years before cold fusion: "If one doubts the necessity for controls, reflect on the statement: 'It has been conclusively demonstrated by hundreds of experiments that the beating of tom-toms will restore the sun after an eclipse.' "

Laura Garwin, who is Dick Garwin's daughter, was an editor at *Nature* and would oversee Pons and Fleischmann's submission to that journal. She said of the *JEAC* paper: "I was extremely surprised that that paper got published. It didn't have the elementary control experiment. The obvious thing you see when you look at that paper is, Why didn't they do it with H_2O? Any high school student could've refereed it, because of that obvious ingredient of the scientific method." Nate Lewis said, "You read the paper and you assume that there are controls done, and they did these things by rational means. Then you find out it wasn't always that way." The absence of controls would haunt the cold fusion episode like a recurring nightmare.

Also conspicuously absent from the Pons-Fleischmann *JEAC* paper was raw data of any kind. All the numbers seemed to have passed through at least one level of interpretation and analysis. It was difficult to determine exactly what that analysis had been, and thus what the original data had been.

On the one hand was the evidence for the by-products of a nuclear reaction—tritium, neutrons, and gamma rays. Even if one believed what Pons and Fleischmann reported in the paper, the numbers were so small as to be almost irrelevant.[24] "As soon as you read the paper," remarked Larry Faulkner of the University of Illinois, "you realize the only thing they'd ever done was calorimetry." Faulkner's point was that even if they had the tritium, neutron, and gamma ray work right, "the radiation was tremendously small compared to the heat effects." Fleischmann himself was now telling inquirers that their observations stood or fell on the production of heat. He didn't understand, he said, why everyone was off looking for the production of neutrons. He called it "a very barren search."

Fleischmann may have been right, but it was still virtually impossible to judge just how much heat Pons and Fleischmann's cells had produced. The paper contained two tables that itemized the amount of heat produced by the cells. All the numbers were given as a rate of "excess heating." In table 1, excess heating was set down as a function of the current density and electrode size. In table 2, which was the more awe-inspiring, excess heating was recorded as a percentage of break even, that being the point at which the energy going to charge and run the cell was equal to the energy, in the form of heat, produced by the cell.

Of the three data columns of table 2, the third was the impressive one, although its meaning came through with less than crystal clarity. Here

Table 2

Generation of excess enthalpy in Pd rod cathodes expressed as a percentage of break-even values. All percentages are based on $^2D + {^2D}$ reactions, i.e., no projection to $^2D + {^3T}$ reactions

Electrode Type	Dimensions /cm	Current Density /mA cm^{-2}	Excess Heating/ % of Break-Even		
			a	b	c
Rods	0.1 × 10	8	23	12	60
		64	19	11	79
		512	5	5	81
	0.2 × 10	8	62	27	286
		64	46	29	247
		512	14	11	189
	0.4 × 10	8	111	53	1224
		64	66	45	438
		512	59	48	839

[a] % of break-even based on Joule heat supplied to cell and anode reaction $4\,OD^- \rightarrow 2\,D_2O + O_2 + 4\,e^-$.

[b] % of break-even based on total energy supplied to cell and anode reaction $4\,OD^- \rightarrow 2\,D_2O + O_2 + 4\,e^-$.

[c] % of break-even based on total energy supplied to cell and for an electrode reaction $D_2 + 2\,OD^- \rightarrow 2\,D_2O + 4\,e^-$ with a cell potential of 0.5 V.

were the tremendous numbers for excess heating: 1224 percent of break even, 839 percent, 438 percent, 286 percent, and so on.

This was not raw data, which is an interesting term if one gives it a moment's thought. *Raw data* has no particular positive or negative connotations. Yet it can be argued that if data is not "raw" then it is "cooked," which does carry negative connotations. So be it. Scientists might have been better served had Pons and Fleischmann chosen to provide the raw data rather than this "interpreted" data. It would take some scientists weeks to figure out where these numbers came from; others, apparently, never would. One assumed that somehow, somewhere Pons and Fleischmann had measured these numbers. The paper didn't seem to say.[25]

The Wrighton-Parker cold fusion collaborators at MIT read the paper simultaneously. As Dave Albagli told it, their unanimous and near immediate response to reading it was, Where's the evidence? Why are we doing this? "You hear they've been doing this for five years," Albagli said, "and you say, Why don't they already have it down? Why isn't this just a brief Nobel Prize–winning paper that says it all, very succinctly and very emphatically?"

There were three possible answers to these questions. The first and simplest was that Pons and Fleischmann had no evidence and cold fusion was a canard. The second was that a man of Fleischmann's reputation was not going to make such a mistake, therefore Pons and Fleischmann must have had the evidence, but they had been rushed to publication by the competition from BYU. Thus the paper was poor, but the work, should all the details be known, was not. To imagine that Pons and Fleischmann, especially Fleischmann, had made these claims without doing the definitive experiments, as the paper seemed to indicate, was a stretch. The third possibility, which became accepted around the University of Utah as the gospel, was that Fleischmann and Pons must be hiding something. As Albagli said, "When you get this incomplete information and you give them the benefit of the doubt, you realize they might be trying to buy themselves some time."

The history of science has a surprising number of precedents for this kind of covert business. When Galileo discovered the phases of Venus in December 1610, he claimed his discovery by anagram: *"Haec immatura a me iam frustra leguntur o.y."* When unscrambled this becomes *"Cynthiae figuras aemulatur Mater Amorum,"* which means, with the poetic metaphors translated as well, that Venus emulates the phases of the moon. Galileo, as the sociologist of science Robert Merton put it, was a "seasoned campaigner" when it came to defending his discoveries from claim jumpers.

The precedent on everyone's mind during cold fusion was Paul Chu's 1987 discovery of a compound that remained superconducting at 90 degrees kelvin, which by superconducting standards was hot enough to be remarkable. It was the discovery of high-temperature superconductors that had momentarily made the scientific community willing to believe in miracles. Chu had submitted a paper in the winter of 1987 to *Physical Review Letters* claiming that the magical ingredient in his superconductor was ytterbium when in fact it was yttrium. Later Chu blamed the error on his secretary, who had also, so Chu said, incorrectly transcribed the proportion of ytterbium, or yttrium, or whatever. (As one writer put it, "She must have been having a very bad day.")[26] Thus, when Stan Pons spoke at an American Chemical Society meeting in Dallas two weeks later, one suspicious individual asked if there were any "typographical errors in the paper worth commenting on."

Hugo Rossi, who was dean of the College of Science at Utah, said he began to have the opinion, or had been given the opinion, that Pons and Fleischmann had written their paper so that it would satisfy the requirements for establishing priority but give no more information. "Martin and Stan," Rossi said, "did not want to get into a position of having to compete in research with millions of laboratories."

John Flynn, who was an expert on patent law at the University of Utah, remarked that he was called in by Chase Peterson after the announcement to help shut the barn door, so to speak, on the patent issue. He remembered advising Peterson and his colleagues that the less said about the technical sides of the discovery the better. "If this [had been] in a commercial lab," Flynn said, "the place would have been locked and sealed and those people wouldn't be allowed to talk to anybody." Flynn observed that Pons was a canny businessman, familiar with inventions and patents, and he knew the risks of a public announcement. So it seemed eminently possible at the time that the paper had been purposely written to be obscure.

There was, of course, an older historical precedent for this as well. Alchemists, traditionally, refused to disclose the secret of the philosophers' stone, which would transmute base metal into gold. As Artephius, an alchemist of moderate repute, put it, "Poor idiot! Could you be so simple-minded as to believe that we would teach you clearly and openly the greatest and most important secrets?" Or as the good Reverend Charles Mackay put it in *Extraordinary Popular Delusions and the Madness of Crowds:*

> Unluckily for their own credit, all these gold-makers are in the same predicament; their great secret loses its worth most wonderfully in the telling, and therefore they keep it snugly to themselves. Perhaps they thought that, if every body could transmute metals, gold would be so plentiful that it would be no longer valuable, and that some new art would be requisite to transmute it back again into steel and iron. If so, society is much indebted to them for their forbearance.

In the end, the quality of the paper didn't seem to matter. By the time it appeared, anyone who liked the sound of cold fusion was hooked already. It had become a self-sustaining phenomenon. Who, whether chemist or physicist, having tried the experiment and become caught up in the excitement, could read the paper, suspect it was scientifically bankrupt, and walk away?

24 CAMBRIDGE. GAMMA RAYS

Amid all the talk of heat and neutrons, it was the gamma rays that should have been the definitive proof of cold fusion. Pons and Fleischmann had placed their gamma ray spectrum in figure 1A of the paper, the position traditionally reserved for the strongest evidence in a scientific case. Here, the author is saying, look, this is my proof.

The reaction in question is known as neutron capture on proton. The fusion event emits a neutron. The neutron is captured by the hydrogen in the water around the cell. This creates a deuterium nucleus, which then releases a gamma ray with exactly 2.22 million electron volts (MeV) of energy. It always releases a gamma ray of exactly 2.22 MeV. The universe is that consistent. "It's the first thing you get in first-year quantum mechanics," explained Richard Petrasso, an MIT physicist who should know. "They'll talk about the deuteron binding energy, and it's 2.22 MeV. It's famous and been around for ages, and if you see a gamma ray coming out at 2.22, you know what's taking place."

Counting neutrons, as Petrasso put it, "can mean diddly," but a gamma ray spectrum, that is an art form with a long and established history. Put a prism in front of a light source, as Isaac Newton first did, and the light fans out into its component colors. Put the prism in front of sunlight, and this spectrum is banded by hundreds of dark lines, each of which represents the absorption of the light at exactly one wavelength by a single chemical element or isotope. If the line is there, it means the element is there. All spectra have valleys, which are absorption lines, and peaks, which are the signatures of distinct atomic processes that emit light, which is composed of photons or gamma rays, which are nothing more than high-energy photons. The same lines appear from the flame of a candle or the light of a galaxy across the universe. This is true in Utah, in China, on the moon, or anywhere else.

Petrasso explained this unequivocally: "Each atom has its particularly unique song; each gamma ray is a particular identification. It's like your name, my name, anything. You show a spectrum, and that conveys a tremendous amount of information and it's very unambiguous. These are powerful signatures. . . . And when you show that line at 2.22, it means one thing, it means neutron capture on proton and it means that you're generating neutrons."

And there it was in figure 1A of the paper. Gamma rays at 2.22 MeV (or 2220 KeV in the units preferred by Pons and Fleischmann), a peak that stood up like the Statue of Liberty or the Eiffel Tower. Unmistakable. Unambiguous.

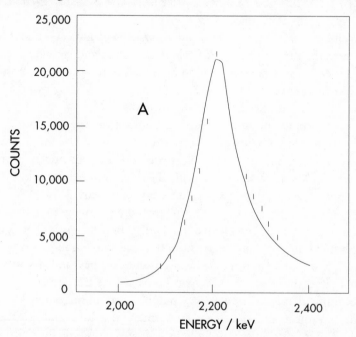

Petrasso looked at that spectrum and found it "pretty startling." And then, as he continued to look at it, he said he grew very disturbed. That a gamma ray spectrum can be disturbing may be the kind of thing that only a serious student of science can appreciate. Nonetheless, these gamma rays became one of the more unfortunate stories in the cold fusion episode.

Petrasso was certainly not the only physicist who noticed that the gamma ray spectrum had its problems, but he had followed up to find out why. "If you look at my career," he said, "you might summarize it by the following: if something makes X rays or gamma rays, I want to know how many, what kind of X rays or gamma rays, and what's behind it."[27]

Talk to enough scientists, and the impression grows that they can be classified into two types: those who went into science because they were good at it, and those who were obsessed with science and had no choice in the matter. Petrasso, at age forty-four, was of the latter variety. He was the son of two musicians, but he said he knew he would be a scientist when he was five years old. Petrasso earned his Ph.D. at Brandeis University and studied, coincidentally, under Ryszard Gajewski. Over the years he had worked on the *Uhuru* satellite, an orbiting X-ray observatory, and then on the X-ray telescope on the Skylab mission. He studied X-ray flares from the sun and for the past decade had been working on X-ray diagnostics for the MIT fusion program, developing techniques for analyzing the plasma in a tokamak by studying the X rays that it emits.

Petrasso considered cold fusion immediately implausible and was unwilling to abandon his faith in nuclear physics on the strength of a press conference, which, he said, "is just glib bullshit, as far as we know." He did not join the Parker-Wrighton cold fusion collaboration because he did not believe cold fusion warranted that kind of effort. Even a year later Petrasso would grow exasperated when discussing cold fusion and especially gamma rays. "For me," he said, "the compelling piece of information is the gamma line. If they've got something, they've got to show me this line. And if they've got it, then, fine, I'll go out and try and reproduce the experiment, but not beforehand."

Once Petrasso glanced at the Pons-Fleischmann gamma ray spectrum, he knew it had serious problems. For starters, the spectrum was clipped, which is to say it was not a spectrum but a single peak, identified at 2.2 MeV, with nothing on either side. It was like showing a single line of color from the spectrum of visible light, except that gamma rays, which are beyond the visible spectrum, are therefore identified not by color but by energy. If Pons and Fleischmann had displayed their entire spectrum, then researchers could have judged for themselves that the gamma ray peak was at 2.22 MeV. Instead, they had to take it as a matter of faith, which was, of course, exasperating.

Why didn't they show the entire spectrum? Petrasso didn't know, but, he said, "in order to get 2.22 without showing the background lines just waves a flag in my face."

Even as an isolated gamma ray peak, the peak Pons and Fleischmann displayed was disturbing. It's a scientific fact that all gamma ray peaks have certain identifying features, the same way faces or anything else have. In particular, they have what are called single and double escape peaks and Compton edges, all of which are properties of the way gamma rays interact in the detectors rather than of the gamma rays themselves. Suffice it to say that when the incoming gamma ray smacks the crystal

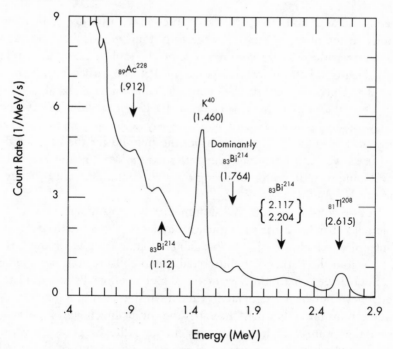

in the detector, it will occasionally skid off to one side, or bounce backward out of the detector, or create new particles, even new gamma rays that might escape from the detector unnoticed. Any of these would result in the detector failing to register some portion of the incoming energy of the gamma ray. Thus, gamma rays of a particular energy show up smeared across a range of lower energies as well. In this smear would be the single and double escape peaks, and the smear would end at the Compton edge. If these peaks and edges "ain't there," as Petrasso said, "it's not a gamma ray."

These characteristics seemed to be lacking in Pons and Fleischmann's single peak, although it was difficult to tell because they hadn't displayed even enough of the background on the low-energy side of the peak to be sure. In the errata to the paper, which appeared shortly after, Pons and Fleischmann still neglected to show the entire spectrum, but they did replot the single peak, revealing slightly more of it on each side. Now it was an inescapable fact that the escape peaks and the Compton edge were not there. Thus, as Petrasso noted, it wasn't a gamma ray peak.

Once the errata appeared, it was possible to measure the width of the new peak. Petrasso did this, as did quite a few other physicists, and realized that the width of the peak was incompatible with the sodium iodide detector Pons and Fleischmann had used. Detectors are like cameras; they can achieve a certain resolution of detail, depending on their quality. The better the detector, the more precisely it can identify the energy of the gamma rays, and the narrower the peak. Pons and Fleischmann's peak was impossibly narrow considering the instrument they had used. So it wasn't a gamma ray peak.

All these defects strongly suggested that the peak was what scientists call an artifact, a bogus signal produced by the equipment in question, which in this case would be the detector. Petrasso considered this an "ironclad" fact.

This then raised the pivotal question: If the peak was an artifact, how did it come to have exactly 2.22 MeV of energy, precisely what would be expected for a sophisticated test tube generating neutrons? If the peak was an artifact, it could have manifested itself at any energy at all. Petrasso found this coincidence very suspicious. "What's the chance of an artifact coming up at precisely the energy 2.22 MeV?" he asked. "It's just about zero."

Lewis Thomas has written that, when scientists are presented with an intellectual challenge, they respond with something very much like aggression: "While it is going on, it looks and feels like aggression: get at it, uncover it, bring it out, grab it it's mine! It is like a primitive running hunt." In this case, it didn't matter that the challenge had nothing to do with adding another brick to the growing edifice of science. Petrasso was affected by the same powerfully hypnotic stimulant: get at it, uncover it. Find the answer. This was further stimulated, one can assume, by his distaste for glib bullshit. "We already knew that the data we were looking at was bogus and was an artifact," he said, "and it's very unnerving because you have to ask yourself, What in the hell is going on here?"

25 DEBRECEN, HUNGARY, AND TOKYO, JAPAN

Observations are useless until they have been interpreted. The analysis of experimental data therefore forms a critical stage in every scientific inquiry—a stage which has been responsible for most of the failures and fallacies of the past.

E. BRIGHT WILSON,
An Introduction to Scientific Research

April Fools' Day. Two Hungarian physicists, Gyula Csikai and Tibor Sztaricskai of Lajos Kossuth University at Debrecen, announced that they had duplicated the electrolysis experiment and detected the emission of neutrons. This was actually a confirmation, albeit premature, of the low-level neutrons reported by Steve Jones and company. Jim Brophy called it the second confirmation of Pons and Fleischmann, intimating, of course, that the first was Jones. A *Wall Street Journal* correspondent interviewed the triumphant Hungarians and reported that "Mr. Sztaricskai was ambiguous about his results, at times saying he had confirmed the Pons-Fleischmann results, and at other times saying he had replicated the lesser, Brigham Young findings." Paul Palmer of BYU was less ambiguous. The Hungarians, he said, "undoubtedly" confirmed the BYU room-temperature fusion discovery.

In Tisbury a British reporter caught up with Martin Fleischmann returning home from his local pub and told him that their experiment had been successfully reproduced. Fleischmann responded, "If it's true, that's fine." It was not true, however. Mr. Sztaricskai and his confirmation soon vanished into the background, never to be heard from again.

Equally premature was the announcement by Professor Noboru Oyama of the Tokyo University of Agriculture and Chemistry. *Nature* reported that Oyama had detected "large amounts of heat and gamma rays" generated by a cold fusion experiment. Apparently Oyama had performed a quick and dirty experiment so he could report results at the April 1 meeting of the Chemical Society of Japan. However, under such time constraints, if less dramatic explanations existed for the experimental results, Oyama had little time to root them out. Other Japanese researchers who had seen the data were not wildly impressed. The *Nature* report continued, "Oyama says he will try to duplicate the experiment with the Japan Atomic Energy Research Institute to determine neutron yield." This sounded like the kind of experimental procedure Oyama might have wished to pursue before reporting his results, but, in the brave new world of cold fusion research, the priorities had changed and therefore so had the rules.

26 SALT LAKE CITY. GROUND ZERO

On April 4, Marvin Hawkins was fired. He found a letter in his school mailbox from Stan Pons, who was lecturing on cold fusion at the University of Indiana. The letter was marked "confidential."

"So I rip it open," said Hawkins, "and to my utter dismay—you could have picked me up with a putty knife off the floor, after having read this—it said he was surprised but relieved that I had returned the remainder of the stolen documents and materials to him. He was glad there didn't have to be any other involvement by the law enforcement agencies or the universities or the lawyers, [but] he was disappointed in my behavior, and he thought it would be appropriate [if] I relinquish my ties with him and the university and return my key to the secretary. And I think, What am I reading here? [I] turn it over and I read it again, just to make sure it wasn't a fantasy or some crazy dream. I'm going, 'I can't believe this.' This is just unreal. This isn't right. I had taken a lot, and I will not be scraped out of this place without a degree."

Hawkins took the letter to public relations and showed it to Barbara Shelley, who had been his contact in the cold fusion work. She seemed sympathetic, and he told her that if the situation was not rectified he'd go to the press with it. He made the same argument to Cheves Walling, the most distinguished chemist at the U and a member of the National Academy of Sciences. Walling said he'd talk to Pons; they'd work it out.

Late that night, when it was already early morning in England, Hawkins called Fleischmann. He recounted the turn of events, and Fleischmann said, "Marvin, don't worry about it. Don't even talk to Stan. Let me talk to Stan. . . . When Stan wants to talk to you, then that's fine. Do not let us down. It's going to make the situation worse."

On the morning of April 5, Pons tracked down Hawkins and they arranged to meet in his office, where they were joined by Walling and Shelley. Apparently, under some pressure, Pons chose to forget the whole thing and told Hawkins to get on with his thesis. Hawkins said one of the conditions he insisted upon at this meeting was that his name be added to the list of authors in the errata, which it was: "M. Fleischmann and S. Pons regret the inadvertent omission of the name of their co-author, Marvin Hawkins. . . ." Few scientists, however, believed that M. Fleischmann and S. Pons had forgotten about Marvin Hawkins in their rush to get out the paper. As Hawkins later said with noticeable bitterness, "When somebody says, 'Oops, we screwed up, we inadvertently omitted one of our coauthors' names and we apologize,' that's absolute bull. Anybody would come across that would realize that's the case."[28] In any case, later Cheves Walling told the story of Marvin Hawkins and the lab books to Chuck Martin of Texas A&M, pretty much as Hawkins told it.

Martin, in fact, was visiting Pons on April 5, the day the lab book issue was finally resolved. Martin had hoped to get details of the experiment from Pons so he could successfully replicate it. "I was frustrated by the fact that I couldn't get at him," Martin said. "He looked like hell. Clearly he hadn't slept in weeks. He was in some weird low gear that I'd never seen in him before, because he was so tired."

As Martin remembered it, Pons told him he couldn't show him the notebooks, because "this kid" had stolen them. This did not jibe with Hawkins's version, in which the books were already returned by this time. Pons did give Martin a cursory tour of the lab and let him briefly glimpse the photocopies, then suggested they go to his house, where they sat and listened to messages on Pons's answering machine. "He got phone calls from the Japanese consulate," said Martin, "from Senator Al Gore; he got phone calls from congressmen, from quacks, from millionaires, from everyone." Again, Martin tried to get Pons to talk about the data and failed.

"Then the strangest thing happened," Martin recalled. At Pons's suggestion, they went off to a hotel, where Pons took to drinking at the bar with a television crew from the Canadian Broadcasting Corporation that was in town putting together a cold fusion documentary. Martin was dumbfounded: "I'm sitting, thinking, Shit, we could be talking science and instead we're sitting at a bar drinking. I figured, Okay, he's under a lot of pressure and maybe he needs this. But it was really strange. Here was Stan in the middle of the day, three or four o'clock, over here drinking with these guys. He wasn't talking science. He was talking to these guys from the CBC, and they were his chums." Martin decided he was wasting his time and took an evening flight back to Texas.

27 THE PRESS

On April 5, *The Wall Street Journal* reported that Brookhaven National Laboratory on Long Island had confirmed the BYU observation of neutrons from cold fusion cells. This was then considered additional confirmation of Pons and Fleischmann, as most of the press and public hadn't differentiated between the two competing groups with their two competing claims. The *Journal* quoted Kelvin Lynn of Brookhaven saying, "We're not absolutely certain, but we have detected neutrons [produced by fusion reactions] that are consistent with the Brigham Young result."

This seemed to lift hopes for the folks back home in Utah. Chase Peterson told *The Deseret News* that Stan Pons had told him, "If I was so sure two weeks ago, I am more sure now."

The *News* also reported that Pons had finally done his control experiment with ordinary water replacing heavy water. "It produced no significant heat," The *News* said. "This could be proof that the heating process is indeed nuclear and not chemical as some physicists have suggested."

Two days later the Associated Press reported that Brookhaven Laboratory had *not*, as previously reported, confirmed the production of neutrons in cold fusion cells. The Brookhaven scientists ran two experiments. One produced nothing, and the other turned up results so small as to be statistically meaningless. Then AP quoted Mark Wrighton of MIT saying that he and his colleagues remained very skeptical. "If nuclear fusion occurs," Wrighton said, "it is at a very low level and our detectors aren't sensitive enough, or it takes longer than ten days, or it doesn't work."

The story then turned to the University of Utah for comment and quoted Pam Fogle saying "that the team believes the reaction can take ten days to three weeks to occur." It had been only two weeks since March 23, so perhaps Fogle was suggesting that Wrighton be patient. On the other hand, nine days before Pons had told Al Bard that the reaction took 7.7 hours to occur.

28 SALT LAKE CITY

"He that doeth nothing is damned, and I don't want to be damned," said Governor Norman Bangerter, shepherding his flock into the promised land. With such apocalyptic oratory, the House and Senate of the state of Utah voted nearly as one—96 yea, 3 nay—to appropriate $5 million to the cause of cold fusion.

"Waiting and seeing," proclaimed W. Eugene Hansen, chairman of the state board of regents, "could mean that the discovery of the century will be developed by Mitsubishi. If there is to be a Fusion Valley, we feel that the Fusion Valley should be here, in the state of Utah."

The legislature proposed that the $5 million be split over two fiscal years, with $2 million for the first, and $3 million for the second. As *The Salt Lake Tribune* reported, this cash would be spent for "equipment, space, retention of consultants, conferences and other 'appropriate' applied science techniques that might lead to market development." The money would only be disbursed after the scientific community provided independent confirmation of the cold fusion findings.

What form this independent confirmation would have to take was not specified. One can assume that on April 7 a convocation of lay politicians believed that such confirmation would be indisputable when it came. Yea or nay. A sign from the heavens. The governor himself would appoint a nine-member panel to decide whether confirmation had been achieved. The panel would consist of "a nuclear physicist, a chemist, an at-large 'member of the scientific community,' the state's science adviser, a certified public accountant, two business-sector members with 'technological research and development' backgrounds and two at-large members of the general public."

And, finally, the legislature proclaimed that this fusion council would be a secret organization; should the people of the great state of Utah want to watch how their cold fusion money was being spent, they were out of luck. This was not, as one might think, a clever ruse by the Utah legislators to protect their reputations should they blunder mightily. Rather, as they later explained, they only wanted to assure that the valuable methods of cold fusion technology could be discussed without fearing that foreign companies would run off with them. Nonetheless, the local Society of Professional Journalists objected, pointing out that the state's Open Meetings Law prohibited such secret business. The governor countered by signing an exemption to the law that would let his fusion council close any meeting in which they expected to discuss intimate technological details. All budget sessions were open, all budget proposals would be public documents, but anyone caught leaking the technology would go to jail.

The only cold fusion proponent in Utah who seemed to be against this kind of support was Stan Pons. Pons seemed to see all the cold fusion boosterism as a misguided effort to hang his reputation as far out on a limb as the citizens of his state would allow. In fact, he had begged Hugo Rossi to stop the $5 million from going through. "I really didn't understand Stan's motivation," Rossi recalled, "and I asked him about it, and he said, 'We don't really know that this is going to work. This is completely premature to be doing this.' " Pons said that Fleischmann had mapped out a research program, complete with their estimated spending for the next eighteen months. He said the money promised from the Department of Energy and the Office of Naval Research would be enough to support them.

Rossi responded that the rationale for the state funds was to keep their research ahead of the rest of the world, that "laboratories with millions of dollars to spend were working on this, and in no time they would be way ahead."

But Pons disagreed. "They can't get way ahead of us," he said. "We have already put five years into it, we have learned lots of things, lot of subleties you have to work through. We'll be fine if we're just left to do our program."

This argument to proceed conservatively had a great deal to recommend it. Pons, however, was no longer in charge. Once the legislature had allocated the $5 million, Pons was left to tell reporters that the state's faith in cold fusion was a "great honor." What else could he say? He added that now they would be able to proceed even faster with their research.

29 SANTA BARBARA, CALIFORNIA

Steve Koonin's initial reaction to cold fusion was that it was crazy: "What are these guys talking about? This can't be right."

But after he thought about it a moment, maybe it wasn't so crazy: "I don't know anything about palladium, but I know it absorbs a lot of hydrogen—I don't know what 'a lot' means, either, [but] maybe in some funny way, some things can happen." The question, of course, was What kinds of things?

Koonin was on sabbatical at UC Santa Barbara when cold fusion broke. He was a professor of theoretical physics at Caltech, where he also ran the Kellogg Laboratory, an experimental physics laboratory. He was a second-generation Lithuanian from Brooklyn and still had the pugnacious demeanor of a New Yorker, looking not unlike an adult version of one of the Dead End Kids. He had done his undergraduate degree at Caltech in the usual four years, although he started when he was sixteen, and obtained his doctorate at MIT in three. He had been back at Caltech ever since. It was Koonin who had supervised the JASON review of Steve Jones's muon-catalyzed fusion.

Koonin initiated his cold fusion research from a neutral position. He perused the literature and simply collected facts. He learned, among other things, about the structure of palladium, the thermodynamics of the hydrogen-palladium system, the time it takes hydrogen to diffuse through a palladium lattice, and the energy required to absorb hydrogen (or deuterium) into the palladium.

This last point was a crucial one. Pons and Fleischmann had claimed to get inexplicable amounts of energy out of their palladium rods. "So," Koonin explained, "you want to know how much energy do I really get out if I just dump all the hydrogen out again? Now that turns out to be a lot. You could explain an awful lot of heat on this deabsorption of hydrogen. There's a lot of energy in there, equivalent to a reasonably high explosive."

Because Koonin had once written a textbook on computational physics, he then decided to use one of its programs to do "the most accurate calculations that one could ever get" on the possibility of two deuterium nuclei fusing. His partner in this endeavor into nuclear theory was Michael Nauenberg, a German-born physicist out of UC Santa Cruz. Nauenberg had grown up in Colombia after his family fled Germany in 1939. He had obtained his doctorate under Hans Bethe at Cornell and in 1966—after stints at the Institute for Advanced Study at Princeton, at Columbia, and at Stanford—had become one of the first faculty members to join the UC Santa Cruz physics department. Over the years Nauenberg had dabbled in condensed matter physics, statistical mechanics, astrophysics, nuclear physics, and dynamical systems, commonly known as chaos theory. Nauenberg was also recharging his batteries in Santa Barbara.